Robotics

Edited by Kenneth Partridge

The Reference Shelf
Volume 82 • Number 1
The H.W. Wilson Company
New York • Dublin
2010

The Reference Shelf

The books in this series contain reprints of articles, excerpts from books, addresses on current issues, and studies of social trends in the United States and other countries. There are six separately bound numbers in each volume, all of which are usually published in the same calendar year. Numbers one through five are each devoted to a single subject, providing background information and discussion from various points of view and concluding with a subject index and comprehensive bibliography that lists books, pamphlets, and abstracts of additional articles on the subject. The final number of each volume is a collection of recent speeches, and it contains a cumulative speaker index. Books in the series may be purchased individually or on subscription.

Library of Congress has cataloged this serial title as follows:

Robotics / edited by Kenneth Partridge.
p. cm.— (The reference shelf ; v. 82, no. 1)
Includes bibliographical references and index.
ISBN 978-0-8242-1096-0 (alk. paper)
1. Robotics. 2. Robots. I. Partridge, Kenneth, 1980-
TJ211.R556 2010
629.8'92—dc22

2009050935

Cover: BERTI the robot interacts with a Sony AIBO robot dog at The Science Museum's Antenna Gallery on February 17, 2009 in London. BERTI is a life size humanoid robot built to mimic a key component of human communication: gesturing. Visitors to the Antenna Gallery will witness BERTI's gestural abilities and get the chance to participate in live research by telling scientists how they rate BERTI's performance. (Photo by Peter Macdiarmid/Getty Images).

Visit H.W. Wilson's Web site: www.hwwilson.com

Printed in the United States of America

The

Reference Shelf®

Contents

Preface

In his 1942 short story "Runaround," the legendary science fiction writer Isaac Asimov outlines his famous "three laws of robotics," a set of principles that governed much of his subsequent work and influenced generations of fiction writers, scientists, engineers, and philosophers. According to Asimov, a robot "may not injure a human being," "must obey any orders given to it by human beings," and "must protect its own existence, as long as such protection does not conflict with the First or Second Law." Asimov sometimes broke these rules, but even so, they speak to a certain idealized notion of robots, suggesting that intelligent, self-aware machines might simultaneously serve and live in harmony with their human creators.

While such optimistic views have proven popular—*The Jetsons*, *Short Circuit*, and *Wall-E* are three of the many movies and television shows that have featured lovable, benevolent robot characters—they're not the only ones that have taken root in the public's consciousness. From 1921's *RUR (Rossum's Universal Robots)*—the Czech play that introduced the term "robot"—to the more recent "Terminator" films and TV series, people have long considered the dangers of "playing god" and creating artificial life. Just as Adam and Eve refused to follow the rules of their creator, robots featured in sci-fi tales often rebel against humans, using superior strength and intellect to take over the world.

One decade into the 21st century, neither vision has come to pass. While the world is home to millions of robots—defined by the Carnegie Science Center as any machine that "gathers information about its environment (senses) and uses that information (thinks) to follow instructions to do work (acts)"—few bear any resemblance to the walking, talking humanoids so prevalent in popular culture. Most are decidedly less glamorous: assembly-line arms, Roomba vacuums, pill-sorting and supply-schlepping hospital helpers, and military "drone" flyers. To be sure, these and other real-life robots represent stunning technological breakthroughs, but none are advanced enough to pass for human.

Nevertheless, this issue of The Reference Shelf arrives at a pivotal time, as scientists edge ever closer to fulfilling the fantasies of sci-fi enthusiasts and creating robots that—thanks to either expert programming or the development of genuine "artificial intelligence" (AI)—have the power to operate autonomously and interact with people. This volume aims to present a brief history of robotics, examine how far the field has progressed, and look to the future, considering both the promises and perils of building increasingly complex robots.

The first chapter, "Of Men and Machines: Robotics—Past, Present, and Future," contains articles that place today's robots in historical context. The pieces also consider how different cultures view these technological marvels, contrasting the attitudes of the Japanese—who, according to the author of one entry, have largely embraced them—with people from other nations, who have expressed more trepidation, fearing that, among their other downsides, mechanized workers are rendering human ones obsolete.

Selections in the second chapter, "Faster, Stronger, Cheaper: Robotic Automation in the Workplace," take a closer look at what effects robots have had on a variety of industries. Some experts contend that robots stand to increase productivity without costing people jobs, while others say "othersourcing"—the practice of replacing humans with machines—is a serious problem that will only get worse. The next chapter, "The Robot Will See You Now: Robotics in Health Care," features selections that center on how machines are helping doctors and health-care professionals better serve patients. As if the da Vinci Surgical System, a device that allows surgeons to operate via remote control and make remarkably small incisions, weren't amazing enough, scientists have now set their sights on building microbots and nanobots—machines tiny enough to swim through patients' bodies and fight disease from the inside.

Moving from the inner workings of the human body to the furthest regions of outer space, pieces in the fourth chapter, "Robots in Space," detail efforts to explore the cosmos using unmanned spacecraft. Already, the National Aeronautics and Space Administration (NASA) Mars rovers Spirit and Opportunity are bolstering scientists' knowledge of the Red Planet, and in the coming decades, robots might be used to build lunar space stations and relay information from beyond our solar system.

NASA isn't the only government agency with an interest in robots. "Lifesavers or Killing Machines? The Pros and Cons of Robots in the Military," the next chapter, comprises articles that look at how machines are replacing men on the battlefield. The selections pay particular attention to the moral questions associated with using robots as weapons. Entries in the final chapter, "Thinking, Feeling Robots: The Dream (or Nightmare) of Artificial Intelligence," explore the feasibility and philosophical implications of building truly human-like robots. While some doubt software or circuitry will ever give robots brains that rival man's, others believe the line between human and machine is destined to be erased.

Robots may eventually take over, but they haven't yet, and I'm happy to report that this book was made with the help of real, live humans. I'd like to express my sincerest gratitude to all of the writers who allowed their work to be reprinted herein, as well as to my colleagues Paul McCaffrey and Richard Stein, editors whose wit and intellect scientists would be hard-pressed to duplicate. I would also like to thank my parents, Deb and Ken Sr., and my new wife, Lindsey. Don't worry, guys: Should the robots attacks, I'll be ready!

Kenneth Partridge
February 2010

1

Of Men and Machines: Robotics—Past, Present, and Future

Editor's Introduction

Humans have long approached robotics with a mixture of caution and wonder. The term "robot" dates back to 1921, the year Czech playwright Karel Capek premiered the science-fiction landmark *RUR (Rossum's Universal Robots).* In the play, subservient humanoid robots—their name derived from the Czech word "robota," meaning "tedious labor"—come to resent their role as slaves and rebel against their creators. The revolt leads to the eradication of mankind, and even though the robots eventually develop human characteristics, such as the ability to love, *RUR* presents a dark vision of the future, illustrating the potentially disastrous consequences of creating artificial beings.

Nearly a century later, the issues central to Capek's story are more relevant—and bedeviling—than ever. While some would view recent advances in factory automation and the growing popularity of the floor-cleaning Roomba as harbingers of exciting things to come, others worry that scientists and engineers are leading humanity into dangerous territory. The articles in this chapter offer a brief history of man's relationship with robots, highlighting the ways in which machines stand to improve our lives, as well as the myriad problems that could arise if robots learn to think, feel, and act for themselves.

In "Robots May Storm World—But First, Soccer," writer Mikiko Miyakawa traces a path from ancient Greece to the present, analyzing how our concept of robots has changed over time. "People have long dreamed of creating something to help them in their work," Susumu Tachi, a professor at Tokyo University, tells Miyakawa, outlining what he believes has been one constant, even as robots have evolved from simple machines designed to carry out repetitive motions to humanoids capable of speaking and recognizing faces.

In "Better than People: Japan's Humanoid Robots," the second entry in this chapter, a writer for *The Economist* considers why the Japanese have been so quick to embrace robots compared to other cultures. As the country grapples with an aging population and a shortage of young workers, the author observes, many Japanese have shown a preference for filling the resulting labor gap with robots rather than immigrant workers. The writer posits that this attitude has to do with the nation's native religion, Shintoism, as well as the positive portrayal of robots in popular culture.

The third piece, "The Robotic Economy: Brave New World or a Return to

Slavery?" finds Arnold Brown exploring the phenomenon of "othersourcing," or using robots to replace human workers. According to researcher Neil MacDonald, othersourcing in the information-technology sector will, over the next decade, eliminate ten times more jobs than outsourcing, a practice that receives far more attention in the media.

Alan S. Brown opens "Robot Population Explosion," the fourth selection in this chapter, with an eye-popping prediction: "By the end of 2011 the world's population of service robots could exceed the population of Chile—more than 17 million units." While Brown admits the development of complex humanoid machines remains a long way off, he discusses the various ways companies are using existing robot technology to cut costs and increase productivity.

In the fifth and final piece, "Science Reaches for the Dream Machine," writer Conrad Walter looks at the difficulties associated with creating truly self-aware humanoid robots. Speaking critically of Honda's Asimo, a humanoid machine that is able to, among other simple tasks, kick a soccer ball, robotics professor Hugh Durrant-Whyte tells Walter, "If you gave it a different ball, you can forget it." Walters also discusses the concept of "robotic swarms," or groups of machines that work in tandem to perform certain duties. "If you get the balance right," professor Ray Jarvis says, "you have the benefit of some coordination but a certain amount of autonomy that gives you higher reliability."

Robots May Storm World—But First, Soccer*

By Mikiko Miyakawa
The Daily Yomiuri (Tokyo), January 1, 2003

Robots pose no threat to Ronaldo—yet. But in 2050, a soccer team of humanoid robots may be able to beat the human World Cup champion team.

Chasing this seemingly reckless goal, a group of Japanese scientists in 1997 launched RoboCup, the robotic soccer world championship. From these humble beginnings, the annual event has exceeded the organizers' initial expectations.

"The project started with 31 teams from 10 countries, but it rose to as many as 200 from 30 countries in 2002," said RoboCup Federation President Minoru Asada, a professor at Osaka University.

But as well as seeing the size of the event blossom, organizers—and the field of robot research—have seen dramatic developments in robotics, such as the omnidirectional vision and movement that enables robot players to see and move around the pitch much more efficiently.

In June last year, while Japan and South Korea were busy cohosting the human soccer World Cup finals, RoboCup 2002 also was held in the two countries, in the cities of Fukuoka and Pusan.

The robot players competed in four leagues—small, midsize, Sony four-legged (Aibo) and humanoid. The event also featured simulated soccer, rescue competitions and the RoboCup Junior competition for children. But the crowd favorites were the Aibo and humanoid leagues, Asada said.

In the Aibo league, some of the robot dogs actually celebrated their teammates' goals by standing on their hind legs—a sign of communication between the robots.

While the ultimate goal of RoboCup is to reach a point where robot teams can beat human players, the event is expected to spawn useful technologies as designers of the robot players strive toward that goal, according to Asada.

Asada likened the situation to the major impact the space race of the 1960s had on society. That the U.S. National Aeronautics and Space Administration won the space race with the Apollo moon landing was not as significant in the wider scheme of things as the multitude of technologies that were generated in the race to the moon—technologies that would have far-reaching effects in our daily lives.

"RoboCup opened up a new horizon for multirobot research," Asada said. One application of multiagent system research will be in rescue operations where a squad of robots would be able to work together more efficiently than if each robot worked alone.

The event is also aimed at developing robotics that work in harmony with human beings, Asada said.

"Robotics research actually involves all sorts of studies on humans," he said. "In this regard, a robot is an artifact that mirrors people."

Asada is confident that Japan is and will remain the forerunner in the field of robotics.

"The government is trying to promote information technology and biotechnology, but I believe robotics is the industry that will save the country in the 21st century," he said.

EMBEDDED IN OUR UNCONSCIOUS

The history of robots dates back almost 2,800 years. In the ancient Greek epic the "Iliad," written in the 8th century B.C., Homer depicted what is believed to be three prototype robots—a robot that moved around on wheels, a humanoid robot and a robot designed to work in a factory.

"People have long dreamed of creating something to help them in their work," Tokyo University Prof. Susumu Tachi, an expert in robotics, said.

The term itself comes from the Czech word robota, meaning drudgery or servitude—a robotnik is a serf—and was first used in the 1920 play "R.U.R." (Rossum's Universal Robots), by Czech author Karel Capek. "R.U.R." depicts a society that has become completely dependent on mechanical workers capable of doing any kind of mental or physical work.

The play takes on a tragic dimension when the robots develop humanlike feelings and begin to rebel against their human masters, eventually slaughtering them and starting a robot nation.

Tachi said the play was a cautionary tale for scientists about the ethics of developing robots and technology in general.

He cited the Three Laws of Robotics written by the late U.S. science fiction writer Isaac Asimov in 1950:

- A robot may not injure a human being, or, through inaction, allow a human being to come to harm.
- A robot must obey the orders given to it by human beings except where such orders would conflict with the First Law.

- A robot must protect its own existence as long as such protection does not conflict with the First or Second Law.

While the laws have their origin as a literary device, "as robotics became more advanced, people began to realize the significance of the laws," the professor said.

In the 18th century during the Edo period (1603–1868), karakuri ningyo, or windup dolls, were developed in Japan. Osaka University's Asada said the dolls were early Japanese robot prototypes.

But it was only after World War II that robots had any practical use. In 1960, Joseph Engelberger developed Unimate, the world's first robot with a real-life function. Simplistic by today's standards, this first-generation robot simply did the same thing over and over again.

Kawasaki Heavy Industries, Ltd. bought Unimate technology from Engelberger's firm, and in 1969, Unimate robots took to the factory floors for the first time in Japan.

In 1980, three out of every five robots worldwide were in Japan, earning the country the nickname "robot kingdom." Since then, Japan has led the world in this field.

Second-generation robots, also called sensor robots, used built-in sensors to collect data on their surroundings.

Third-generation robots, which first appeared in the 1970s, could move autonomously, and found uses in maintenance and inspection work in seabed oil fields and nuclear power plants.

In 1996, Honda Motor Co. developed P2, a humanoid robot capable of walking on two legs and a predecessor of Asimo. In 1999, Sony Corp. released its first Aibo pet robot.

With these developments, Tachi said, robotics has entered its fourth generation, in which human beings and robots coexist and cooperate. "Initially, robots worked in factories and other limited spaces, but they gradually began to be seen in other places such as industrial complexes, and now they have begun entering households," Tachi said.

Tachi is currently involved in the Humanoid Robotics Project (HRP), a five-year program initiated by the Economy, Trade and Industry Ministry. Through technology dubbed real-time remote robotics, robots will be able to relay sensory data back to a distant human, enabling people to "see, hear and feel" as if they were actually standing in the robot's place, according to Tachi. HRP's initial results will be made public in March.

FIRST OUR HEARTS, THEN OUR MINDS

According to the Japan Robot Association, the market for industrial robots was worth 400 billion yen in 2001. If its predictions are sound and robots, including entertainment robots, are bought by more households, then the domestic robot market will expand to 3 trillion yen in 2010 and 8 trillion yen in 2025.

"Just as almost every household has a computer, we're assuming every household will have a robot," an official of the association said.

The year of 2003 will be "the year of the robot," said Kenji Kimura, president of the Business Design Laboratory in Nagoya, which is planning to launch the world's first robot "that can communicate with people by recognizing their feelings."

Kimura went on to say, "This invention is significant in that our robot understands and expresses feelings."

Certainly, the robot can understand simple questions. Ask its age, and the 40-centimeter-tall, five-kilogram robot tentatively named If replies: "Five years old." But complicated questions confuse If, who replies "I don't understand," and assumes an embarrassed facial expression.

It can distinguish the faces of up to 15 people, "learning" their body language and speech habits.

If's debut is planned for April 7—an important date for robot watchers, as it is the birthday of Tetsuwan Atomu (Astro Boy), the robot boy created by Osamu Tezuka.

The sociable robot, which so far has cost 460 million yen to develop, is able to communicate naturally thanks to the Sensibility Technology software engine developed by the Tokyo-based company AGI. The software "reads" the flow of conversation and makes intelligent guesses about speakers' emotions, according to AGI.

But the robot has yet to undergo a critical step—in February, its software and hardware must be joined.

Kimura, who also serves as the director general of the Human Robot Consortium, a joint venture established by the public and private sectors and universities for the project, hopes that robotics will be as lucrative an industry for the consortium members as the auto industry proved to be for Japan's big carmakers.

Better than People*

Japan's Humanoid Robots

The Economist, December 24, 2005

Her name is MARIE, and her impressive set of skills comes in handy in a nursing home. MARIE can walk around under her own power. She can distinguish among similar-looking objects, such as different bottles of medicine, and has a delicate enough touch to work with frail patients. MARIE can interpret a range of facial expressions and gestures, and respond in ways that suggest compassion. Although her language skills are not ideal, she can recognise speech and respond clearly. Above all, she is inexpensive. Unfortunately for MARIE, however, she has one glaring trait that makes it hard for Japanese patients to accept her: she is a flesh-and-blood human being from the Philippines. If only she were a robot instead.

Robots, you see, are wonderful creatures, as many a Japanese will tell you. They are getting more adept all the time, and before too long will be able to do cheaply and easily many tasks that human workers do now. They will care for the sick, collect the rubbish, guard homes and offices, and give directions on the street.

This is great news in Japan, where the population has peaked, and may have begun shrinking in 2005. With too few young workers supporting an ageing population, somebody—or something—needs to fill the gap, especially since many of Japan's young people will be needed in science, business and other creative or knowledge-intensive jobs.

Many workers from low-wage countries are eager to work in Japan. The Philippines, for example, has over 350,000 trained nurses, and has been pleading with Japan—which accepts only a token few—to let more in. Foreign pundits keep telling Japan to do itself a favour and make better use of cheap imported labour. But the consensus among Japanese is that visions of a future in which immigrant workers live harmoniously and unobtrusively in Japan are pure fancy. Making humanoid robots is clearly the simple and practical way to go.

Japan certainly has the technology. It is already the world leader in making industrial robots, which look nothing like pets or people but increasingly do much of the work in its factories. Japan is also racing far ahead of other countries in developing robots with more human features, or that can interact more easily with people. A government report released this May estimated that the market for "service robots" will reach ¥1.1 trillion ($10 billion) within a decade.

The country showed off its newest robots at a world exposition this summer in Aichi prefecture. More than 22m visitors came, 95% of them Japanese. The robots stole the show, from the nanny robot that babysits to a Toyota that plays a trumpet. And Japan's robots do not confine their talents to controlled environments. As they gain skills and confidence, robots such as Sony's QRIO (pronounced "curio") and Honda's ASIMO are venturing to unlikely places. They have attended factory openings, greeted foreign leaders, and rung the opening bell on the New York Stock Exchange. ASIMO can even take the stage to accept awards.

So Japan will need workers, and it is learning how to make robots that can do many of their jobs. But the country's keen interest in robots may also reflect something else: it seems that plenty of Japanese really like dealing with robots.

Few Japanese have the fear of robots that seems to haunt westerners in seminars and Hollywood films. In western popular culture, robots are often a threat, either because they are manipulated by sinister forces or because something goes horribly wrong with them. By contrast, most Japanese view robots as friendly and benign. Robots like people, and can do good.

The Japanese are well aware of this cultural divide, and commentators devote lots of attention to explaining it. The two most favoured theories, which are assumed to reinforce each other, involve religion and popular culture.

Most Japanese take an eclectic approach to religious beliefs, and the native religion, Shintoism, is infused with animism: it does not make clear distinctions between inanimate things and organic beings. A popular Japanese theory about robots, therefore, is that there is no need to explain why Japanese are fond of them: what needs explaining, rather, is why westerners allow their Christian hang-ups to get in the way of a good technology. When Honda started making real progress with its humanoid-robot project, it consulted the Vatican on whether westerners would object to a robot made in man's image.

Japanese popular culture has also consistently portrayed robots in a positive light, ever since Japan created its first famous cartoon robot, Tetsuwan Atomu, in 1951. Its name in Japanese refers to its atomic heart. Putting a nuclear core into a cartoon robot less than a decade after Hiroshima and Nagasaki might seem an odd way to endear people to the new character. But Tetsuwan Atomu—being a robot, rather than a human—was able to use the technology for good.

Over the past half century, scores of other Japanese cartoons and films have featured benign robots that work with humans, in some cases even blending with them. One of the latest is a film called "Hinokio," in which a reclusive boy sends a robot to school on his behalf and uses virtual-reality technology to interact with classmates. Among the broad Japanese public, it is a short leap to hope that real-world robots will soon be able to pursue good causes, whether helping to detect landmines in war-zones or finding and rescuing victims of disasters.

The prevailing view in Japan is that the country is lucky to be uninhibited by robophobia. With fewer of the complexes that trouble many westerners, so the theory goes, Japan is free to make use of a great new tool, just when its needs and abilities are happily about to converge. "Of all the nations involved in such research," the Japan Times wrote in a 2004 editorial, "Japan is the most inclined to approach it in a spirit of fun."

These sanguine explanations, however, may capture only part of the story. Although they are at ease with robots, many Japanese are not as comfortable around other people. That is especially true of foreigners. Immigrants cannot be programmed as robots can. You never know when they will do something spontaneous, ask an awkward question, or use the wrong honorific in conversation. But, even leaving foreigners out of it, being Japanese, and having always to watch what you say and do around others, is no picnic.

It is no surprise, therefore, that Japanese researchers are forging ahead with research on human interfaces. For many jobs, after all, lifelike features are superfluous. A robotic arm can gently help to lift and reposition hospital patients without being attached to a humanoid form. The same goes for robotic spoons that make it easier for the infirm to feed themselves, power suits that help lift heavy grocery bags, and a variety of machines that watch the house, vacuum the carpet and so on. Yet the demand for better robots in Japan goes far beyond such functionality. Many Japanese seem to like robot versions of living creatures precisely because they are different from the real thing.

An obvious example is AIBO, the robotic dog that Sony began selling in 1999. The bulk of its sales have been in Japan, and the company says there is a big difference between Japanese and American consumers. American AIBO buyers tend to be computer geeks who want to hack the robotic dog's programming and delve in its innards. Most Japanese consumers, by contrast, like AIBO because it is a clean, safe and predictable pet.

AIBO is just a fake dog. As the country gets better at building interactive robots, their advantages for Japanese users will multiply. Hiroshi Ishiguro, a robotocist at Osaka University, cites the example of asking directions. In Japan, says Mr Ishiguro, people are even more reluctant than in other places to approach a stranger. Building robotic traffic police and guides will make it easier for people to overcome their diffidence.

Karl MacDorman, another researcher at Osaka, sees similar social forces at work. Interacting with other people can be difficult for the Japanese, he says, "because they always have to think about what the other person is feeling, and how what they say will affect the other person." But it is impossible to embarrass a robot, or be embarrassed, by saying the wrong thing.

To understand how Japanese might find robots less intimidating than people, Mr MacDorman has been investigating eye movements, using headsets that monitor where subjects are looking. One oft-cited myth about Japanese, that they rarely make eye contact, is not really true. When answering questions put by another Japanese, Mr MacDorman's subjects made eye contact around 30% of the time. But Japanese subjects behave intriguingly when they talk to Mr Ishiguro's android, ReplieeQ1. The android's face has been modeled on that of a famous newsreader, and sophisticated actuators allow it to mimic her facial movements. When answering the android's questions, Mr MacDorman's Japanese subjects were much more likely to look it in the eye than they were a real person. Mr MacDorman wants to do more tests, but he surmises that the discomfort many Japanese feel when dealing with other people has something to do with his results, and that they are much more at ease when talking to an android.

Eventually, interactive robots are going to become more common, not just in Japan but in other rich countries as well. As children and the elderly begin spending time with them, they are likely to develop emotional reactions to such lifelike machines. That is human nature. Upon meeting Sony's QRIO, your correspondent promptly referred to it as "him" three times, despite trying to remember that it is just a battery-operated device.

What seems to set Japan apart from other countries is that few Japanese are all that worried about the effects that hordes of robots might have on its citizens. Nobody seems prepared to ask awkward questions about how it might turn out. If this bold social experiment produces lots of isolated people, there will of course be an outlet for their loneliness: they can confide in their robot pets and partners. Only in Japan could this be thought less risky than having a compassionate Filipina drop by for a chat.

The Robotic Economy*

Brave New World or a Return to Slavery?

By Arnold Brown
The Futurist, July 2006/August 2006

A world run by robots is no longer a notion exclusive to science fiction. The first glimmers of the coming robotic era are already visible.

There are now more than 1.5 million robot vacuum cleaners in use. Robot rovers explore the surface of Mars. Microsoft has created robot teddy bears capable of monitoring kids. Another Microsoft robot, SmartPhlow, will monitor and control traffic flows. There are now even robot camel jockeys in the Middle East.

Signs of the growing prevalence of robotic technology are all around us. But we have yet to fully explore the consequences of our increasing dependence on these machines and the numerous ways they are inserting themselves into our daily lives.

The mechanical slaves of the twenty-first century will perform tasks deemed too hazardous for humans, such as cleaning up toxic waste. Others, deliberately made to resemble humans, will be companions and teachers of children. Some will even be chimeras made up partly of human cells. And, increasingly, they will be both self-repairing and self-reproducing.

If you discard the word "robot," which was coined by Karl Capek in the play *R.U.R.*, you might find that the most apt term for the machines that will increasingly do our manual labor, operate and direct interactions between people and institutions, perform domestic services, fight our wars, take care of children and seniors, clean up our messes, and so on may be slaves.

Robots are not free. They are owned. As of now, they enjoy none of the rights we associate with free human beings. At the same time, our economic prosperity (and much else) is dependent not only on the competency of these increasingly intelligent devices and systems, but also on their compliance.

* Originally published in *The Futurist*. Used with permission from the World Future Society, 7910 Woodmont Avenue, Suite 450, Bethesda, Maryland 20814 USA. Telephone: 301-656-8274; www.wfs.org

JAPAN LEADS THE WAY

Much of the impetus for robot development comes from Japan, where demographic trends and labor costs are creating a growing market for machines that can replace humans. Hitachi's robot EMIEW can be trained to do any number of factory and office jobs. Virtual pets, such as Neapets, are astoundingly popular in Japan, and their popularity is spreading. One robot toy—Pleo, a dinosaur—is designed to elicit emotional responses from children and adjust its own behavior in turn. A group of Japanese scientists has invented a soccer-playing robot called ViSiON; they claim a team of such robots will win the World Cup by 2050. Japanese engineers are rushing to produce humanoid robots to care for the aged as well as children. Also in development are robots that can monitor and assist the elderly in taking medications and help blind people navigate and shop in grocery stores.

The South Korean government intends to roboticize that country, based on a vision of a robot-centered intelligent society.

The U.S. military is another major supporter of robotics. The Pentagon is developing managed "trauma pods" to perform battlefield surgery and plans to spend more than $120 billion to develop what will eventually become autonomous robot soldiers.

As these machines become more humanoid—in appearance, in personality, in thinking—how will their relationship to humans develop? The proliferation of these robots will surely generate much controversy as society ponders what such machines might be legally entitled to. One can imagine a fair amount of resistance to the notion of extending robots (or semi-human robots) the same rights afforded to people, such as the right to own property, vote, or run for office. But discrimination against bionic or semi-bionic entities may be difficult to defend on legal grounds, especially as growing numbers of humans incorporate machinery into their biological functioning.

"OTHERSOURCING" HUMAN JOBS

Much of the sound and fury in U.S. politics in the 2004 election arose from the volatile issues of outsourcing—the movement of work to other entities and other places. The most contentious part of that issue was offshoring, which refers to moving work and jobs to other countries. This particular issue is now creating concern in Europe, as well.

As the effort to break down work processes into components progresses, more and more knowledge work can be outsourced to countries with a lower living wage than places like the United States. As is usual with political sound and fury, the real underlying issue has been obscured by all the noise. That real issue is othersourcing—the increasing ability to have work done not only off-site and by other entities (such as unanticipated competitors) but by nonhumans.

"Over the next 10 years, the rate of IT job loss that can be attributed to automation will be about double what we think will be due to outsourcing," says Neil MacDonald, analyst of the research and advisory company Gartner.

The firm or organization as we have always known it has been a job-creating mechanism. Even when machines have displaced people, the effect has been temporary, because the ultimate outcome of technology has been more jobs. But as work has become more abstract and impersonal—as we have gone from making and growing things to marketing and servicing them—it has become easier to depersonalize work and to see workers as abstract, too.

This trend, together with the extraordinary technological advances of recent years, is leading to a potentially massive shift of increasingly higher kinds of work to machines and software. It has also become easier for unexpected competitors to create a cost or efficiency advantage that can gain them all or part of a business function. This othersourcing shift is seen not only in every kind of business, but also in government, military, and nonprofit activities:

- Computer software models direct the allocation of pharmaceutical company representatives' time and determine the prioritization of customers.
- In many business operations, there is so much information that only computer programs, cellular automata, can process it.
- In many areas of science, experts believe that, because of the overwhelming volume of data, only robot scientists will be able to process it.
- Automated software can cut gemstones as well as or better than humans.
- In the future, airplanes will communicate directly with each other, eliminating the need for human air traffic controllers.
- NASA is developing nanobots for "autonomous nanotechnology swarms" that will know when and how to form shapes and patterns for planetary exploration.
- Robots are being taught to work as teams, using common intelligence.

According to the Robotics Industries Association, sales of factory robots increased 28% in 2005, which comes on top of a 20% rise in 2004. By some estimates, the international market for robotic manufacturing units is already $5 billion.

FROM HUMANKIND TO MINDKIND

In the world of employment, resources have already begun to shift thanks not so much to robotics as to better communications technology. Business leaders have begun to see a growing distinction between right-brain skills and left-brain skills, and they are finding new ways to procure and leverage each. Othersourcing is now extending to many more white-collar jobs where the brainwork is becoming, or has already become, commoditized—even in the field of medicine and drugs. For example, you can buy software to do your taxes instead of having to go to an accountant. This knowledge commoditization will be exacerbated as higher education and technical capabilities expand in lesser developed, lower-wage countries.

Creative work may seem like precisely the sort of area where human workers will remain dominant. But technology is affecting jobs here as well. Instead of hiring whole people to do creative work, managers at many firms are shifting to hiring minds—which don't necessarily have to be human. Eli Lilly's e-business venture, InnoCentive, has more than 85,000 registered "solvers" from 173 countries. Corporations post their biology and chemistry needs on the InnoCentive Web site, hoping that one of the registered researchers will be able to provide a solution.

Similarly, office supply giant Staples held a competition and received 8,300 submissions from customers who came up with new product ideas. Automaker BMW accessed customer creativity by allowing people to post on the company Web site their suggestions for ways to leverage advance telematics and in-car online services. Meanwhile, the BBC has announced its "Creative Archive License," providing public access to its full media archives so that individuals can participate in the production of entertainment.

Initiatives such as these reduce the need for bricks and mortar to house bodies, and all the overhead that comes with hiring people, while expanding the universe of potentially harnessable minds, whatever form they come in. Along with this, the focus will begin to shift away from managing people and toward project management—putting together all the varied resources and components, wherever available and in whatever form, to accomplish the desired task or vision. As organizational energy input continues to migrate away from labor, and organizations that do depend on labor seek out the lowest-cost providers, the management of labor will be less important and the processing and management of information will take center stage.

At some point, even human minds may be superfluous. IBM, École Polytechnique de Lausanne, and others are working toward computer models of the human brain—the key step in designing a computer that will "think" on a par with humans. Ray Kurzweil predicts that we will have an artificial brain that can recognize patterns as human brains do by 2020. But we can already see signs of that distant possibility today. IBM consulting uses a software program that chooses and allocates resources for projects better and faster than humans can. Cisco is using software programs to replace humans in human resources, finance, customer service, and other staff areas.

Another major factor fueling these developments is the increasing complexity of the information age. Recent research shows that people cannot efficiently handle complex problems—those with more than four variables—so we will come more and more to rely on machines to do not only what we don't want to do, but also what we can't do.

WORKING WITH ROBOTIC SLAVES

Questions are already being raised in literature, film, and TV shows about the present and future relationship between humans and these new slaves. Years ago the movie *Spartacus* showed the causes and consequences of revolt by human slaves in ancient Rome. Recent movies such as *Blade Runner*; *I, Robot*; and *Terminator* depict similar—and in many ways more frightening—revolts by nonhuman slaves.

Many economies in the past were slave-based. In ancient Greece, such an economy gave rise to an extraordinarily creative society. In other societies, results were less beneficial. In the twenty-first century, a new slave-based economy is developing, and we have to wonder what the consequences of that will be.

Slave owning can have a corrupting effect, and in some ways it can rob both an economy and a society of energy and aspiration. Furthermore, machines, like humans, can be corrupted. Certainly, the persistent and exasperating problems of software serve as a constant reminder of that corruptibility, as does the common excuse from your banker—"the computer is down." It is estimated that at least 1 million infected computers around the world have been formed into "bot nets," controlled from outside and used to steal identities and surreptitiously install spyware or adware. Advances in technology seem always to bring on problems of control. As transistors get smaller, for example, they become less reliable and predictable, as will the robots and other machines containing them.

The idea of human slavery has not disappeared. One recent report says that at least 12.3 million people in the world are in forced labor-bondage. Others report that the number may exceed 25 million, most of whom are in the sex trade or domestic service.

The machines that we're accustomed to—dishwashers, washing machines, alarm clocks, automatic transmissions, etc.—are clearly inanimate objects. But as more of the machines we own are endowed with human qualities, whether in their appearance or in a nascent ability to think, how will we see the relationship? How will they?

For managers of organizations, these are not just idle questions. The relationships in the workplace between people and machines, already somewhat difficult, will become more complex and uncertain. Managers will have to think increasingly of how to get a specific task done rather than who will do it. Human-resource management is developing into human-machine interface management. HR personnel will need new and different training and experience to do their jobs effectively. Customer relations and service will also be entering new ground. As more of the contact between customers goes through machines, we will require a better understanding of the psychology of such relationships and the effects on customer attitudes and behaviors. (Two institutions that have proven to be leaders in this research area are the MIT Media Lab and the HumanComputer Interaction Lab at Wichita State University.)

As a result of the coming advances in robotics, it is likely that we will see a renewed Luddite movement. This second technophobic movement could well be more difficult to deal with than the one involving machine smashers of the nineteenth century. Employers will be under pressure, from within as well as from politicians and the media, to help those whose jobs will be lost. Retraining programs could help ease the pressure and will most likely be seen as good community relations.

One solution, which could supplement unemployment insurance, is trade adjustment insurance. Meanwhile, as staff and management functions become mechanized, employers will need to explore in depth what the consequences will be as more professionals and managers become afraid and angry. Already facing high costs and inconvenience from theft and sabotage, employers should have a comprehensive othersourcing strategy that anticipates the problems that naturally occur with high layoffs. Businesses could form cooperative skill banks so that employees whose particular skills have been rendered redundant by other-sourcing could be made available to other companies that still need them.

The robotic era may also force us to rethink our tendency to measure people's value solely in terms of their economic contributions. Robotics pioneer Hans Moravec has speculated that, by the year 2050, entire corporations will exist with no human employees, or investors, at all. Yet, there still exists a role for humans in the future workplace. Imagination, empathy, and compassion are still the exclusive domain of Homo sapiens, and, in our automated future, business managers should seize the opportunity to rediscover and encourage these traits in their workers as well as in their customers and users.

The bigger challenge may be learning to live with our progeny even as they surpass us in intelligence and capability. As our mechanical and computerized creations perform more of the tasks formerly performed by humans, and as they come to resemble us to a greater degree, we may need to reconsider more than simply how we interact with these devices. We may need to entirely redefine what it means to be human.

KAREL CAPEK AND THE BIRTH OF THE ROBOT

Karel Capek (1890–1938), a Czechoslovakian playwright, is credited as being the first person to put the word "robot" into popular usage. His play R.U.R. (Rossum's Universal Robots) was first produced on the New York stage in 1922.

According to his obituary, which was published in the *New York Times* on December 26, 1938, "He wrote R.U.R. as a genial satire of the mechanical man. The play revolved around the theme that one day the automation representing the technical perfection of the Western civilization would arise and annihilate its creators. Mr. Capek wrote it as a protest against the progress of Americanization and its technological culture in Europe."

MILESTONES IN ROBOTICS

1921—Karel Capek, a Czechoslovakian playwright, makes famous the word "robot" in the play R.U.R. (Rossum's Universal Robots). The word comes from the Czech robota, which means "tedious labor."
1938—Willard Pollard and Harold Roselund of the DeVilbiss Company design the first programmable paint-spraying mechanism.
1942—Isaac Asimov publishes Runaround, in which he defines the Three Laws of Robotics.
1944—Harvard professor Howard Aiken builds the first electronic computer.
1948—Norbert Wiener, a professor at MIT, publishes *Cybernetics or Control* and *Communication in the Animal*, describing the concept of communications and control in electronic, mechanical, and biological systems.
1951—In France, Raymond Goertz designs the first teleoperated articulated arm for the Atomic Energy Commission.
1954—George Devol designs the first programmable robot. Devol coins the term Universal Automation. He will go on to start a company called Unimation.
1959—Marvin Minsky and John McCarthy establish the Artificial Intelligence Laboratory at MIT.
1960—American Machine and Foundry, later known as AMF Corporation, markets the first cylindrical robot, called the Versatran.
1962—General Motors purchases the first industrial robot from Unimation and installs it on a production line.
1968—Kawasaki licenses hydraulic robot design from Unimation and starts production in Japan.
Stanford Research Institute (SRI) builds a mobile robot with vision capability, controlled by a computer the size of a room. They name the robot "Shakey."
1976—NASA puts robot arms on its Viking I and II space probes.
1997—NASA's Mars Pathfinder mission captures the eyes and imagination of the world as Pathfinder lands on Mars and the Sojourner rover robot sends back images of its travels on the distant planet.
2000—Honda debuts its walking ASIMO robot. Sony unveils its Sony Dream Robots at Robodex.
2001—The Space Station Remote Manipulator System (SSRMS), built by MD Robotics of Canada, is successfully launched into orbit and begins operations to complete assembly of the International Space Station.
2004—Robot rovers Spirit and Opportunity touch down on Mars.
2005—Stanley, a driverless car developed at Stanford University, wins the DARPA Grand Challenge.

Sources: Raymond Hui, www.Trueforce. com. "Robot Milestones (extended)," *Business Week* (March 19, 2001)

Robot Population Explosion*

By Alan S. Brown
Mechanical Engineering, February 2009

The robots are coming. In fact, they're almost here, according to forecasts by industry groups.

Based on predictions by the statistical department of the International Federation of Robotics, by the end of 2011 the world's population of service robots could exceed the population of Chile—more than 17 million units. The Japan Robotics Association, meanwhile, predicts a $15 billion market for service robots by 2015.

Granted, some service robots will have just enough intelligence to do the simplest of jobs. An example is the popular iRobot Corp. Roomba, an autonomous vacuum cleaner. Unlike humans, who plan their work to complete it quickly, the Roomba relies on simple algorithms. It cleans in spirals and follows walls. When it bumps into furniture, it changes angles randomly. It moves slowly, but it cleans.

Led by Roomba, sales of robotic vacuum cleaners reached 3.3 million by the end of 2007, according to the IFR, which represents more than 15 trade groups around the world. Meanwhile, iRobot has introduced simple robots to scrub floors, clean pools, sweep home shops, and eject leaves from gutters. Several companies make robotic lawn mowers, whose sales totaled 110,000 units through the end of 2007. Together, robotic vacuum cleaner and lawn mower sales accounted for $1.3 billion.

More sophisticated are entertainment and leisure robots. Lego Mindstorms is a well-known example. It teaches young people fundamentals by letting them program robot behavior on powerful graphical software. This category also includes sophisticated products, like Sony Corp.'s $2,500 Aibo robotic dog, which could see, recognize commands, and learn new behaviors. Sony discontinued production after 200,000 units as a cost-cutting measure (the fate of many early Japanese entertainment robots). Still, entertainment robots had cumulative sales of two million units worth $2 billion through 2007.

There is an entirely different technology for service robots for professional use. They are much smarter and far more flexible and capable. IFR estimates total sales of professional robots through 2007 were 49,000 units worth about $7.8 billion.

Service robots for defense, rescue, and security applications account for 12,000 units, or 25 percent of the total installed base, IFR reported. Many companies compete in this category. iRobot, for example, has sold more than 2,000 PackBot robots for surveillance, identification of hazardous materials, and detecting and disarming roadside bombs. Great Britain's QinetiQ Group plc makes a similar robot, Talon, which can keep pace with a running soldier and negotiate rough terrain.

Field robots, mainly autonomous milking machines, accounted for 20 percent of the professional service market in IFR's survey. They are followed by sophisticated cleaning robots at 12 percent. An example is Skywash from Germany's Putzmeister Werke, which generates motion programs from CAD diagrams and uses sensors to position itself while cleaning aircraft.

Underwater robots, often used to build and maintain offshore oil pipelines, constitute 12 percent of the market. Smaller applications include construction and demolition robots (9 percent), medical robots (9 percent), and mobile robot platforms for general use (7.4 percent). Logistic and inspection systems accounted for the remaining 5.6 percent.

The market for handicap assistance robots is small, but IFR's members expect it to double over the next four years. IFR believes this application will grow quickly, as robotic technology grows more sophisticated and the populations of the wealthiest nations age. It foresees such key developments as assistive robots for disabled and handicapped persons, as well as robotic prostheses.

Overall, IFR projects the world will install 54,000 new professional service robots worth $9 billion between 2008 and 2011. That may not sound like a lot, especially when compared with projected sales of 12.1 million personal service robots over the same period.

Yet high-end personal robots, especially humanoid robots, are growing ever more sophisticated. While general-purpose humanoid robots capable of helping with complex tasks remain a distant goal, several well-known Japanese companies (Honda, Kawada, and Toyota among them) and a handful of Korean and Chinese businesses are developing them. Not only is the world robot population increasing, but it is moving far beyond carpet cleaning.

Science Reaches for the Dream Machine*

By Conrad Walter
Sydney Morning Herald, December 11, 2007

If you crave flawless skin, order it from Texas. For exquisite hands, try Britain. And for the most sensuous of noses, you would be hard-pressed to look beyond Melbourne. If this sounds like the shopping list of a globe-trotting plastic surgeon, discard any thoughts of Hollywood scalpels. These are among the advanced components being developed around the world as scientists edge towards the dream of building the ideal humanoid robot.

Twenty years ago, there were predictions of robots that would sweep our floors, drive our cars and care for the elderly. In fact, the scope of these challenges has proved far harder than many people expected or acknowledged.

But complexity has not thwarted researchers from making advances in robotics intended to address each of those tasks—plus many others. And while not all robots resemble humans, for those that do, the various elements are rapidly coming together.

The skin, developed by Hanson Robotics in Dallas, is made from a patented polymer called Frubber. The material mimics the elasticity of skin but is more pliable than foam rubber and, when coupled with robotic 'muscles', produces very convincing expressions.

The pneumatic hands are custom-made by a group of dreadlocked 21-year-olds, says Peter Hope, the Sydney distributor for the British Shadow Robot Company. A full hand, with 34 tactile sensors in each of its five fingers, sells for about $250,000.

The nose, with the unappealing acronym RAT (reactive autonomous testbed), is the achievement of an associate professor at Monash University, Andrew Russell. His 'smellbot' is not nearly as discriminating as a human nose, but Russell has programmed his electronic snouts to track specified smells.

About 70 experts in robotics, largely from Australia and New Zealand, and

* Article by Conrad Walters, originally published in the *Sydney Morning Herald*.

from Asian universities, are meeting in Brisbane this week for the annual Australasian Conference on Robotics and Automation.

They will be discussing advances in computer vision that enable robots to avoid collisions, artificial intelligence, a 'human body parts detector' and biologically inspired devices such as robotic fish, says Dr Matthew Dunbabin, a senior research scientist at the Autonomous Systems Laboratory within the CSIRO.

Understandably, public fascination with robotics has gravitated towards those devices that resemble ourselves. Honda's Asimo robot, which is touring Australia, has enthralled children, who cheer as it strides through a lightshow worthy of a rock concert.

Billed as the world's most advanced humanoid robot, Asimo can trot across a stage at 6kmh, deliver a cappuccino to a designated spot, kick a soccer ball and execute the staccato moves of, well, robot dancing.

Professor Hugh Durrant-Whyte, who is internationally recognised for his work to help robots cope with the uncertainties they encounter, is impressed by Asimo's technical achievements, although a conversation with him is rather like listening to one magician whisper trade secrets into your ear as another performs.

Durrant-Whyte, the research director at the Australian Centre for Field Robotics at the University of Sydney, says it all comes down to programming and algorithms. 'If you gave it a different ball, you can forget it,' he says of Asimo's soccer prowess.

Likewise, the robot's ability to climb is linked to a particular set of stairs. If anything changes, Asimo topples. And, as videos on the internet gleefully show, even familiar stairs do not guarantee success.

None of this diminishes the accomplishment, the professor believes, but it does show there is a slim distinction between a robot we perceive as moderately self-aware and one that is brilliantly programmed.

Much of the work being done today focuses on commercial robotics, says Durrant-Whyte, who is leading a $21 million project funded by the mining giant Rio Tinto to investigate using robots for mining.

'I think you'll see real robots in applications like mining and agriculture within the next five years,' he says. 'And Australia will probably become one of the biggest users of robots because it just doesn't have enough people to do what it needs.'

Sometimes, though, the work is not about what people can do so much as what robots can do better, or more safely.

In Melbourne, some of Russell's latest work involves his TASTI robots. One of these robots follows a scent by 'licking' the ground as a snake would; another searches out a specific smell, and might be used to track a chemical plume back to its toxic source.

One of Russell's Monash University colleagues, Professor Ray Jarvis, has been looking at new strategies to overcome the core problems in robotics by studying something called robotic swarms.

For decades, scientists focused on making a super-robot, one versatile enough to perform multiple roles around the house or office, but Jarvis says a promising alternative strategy involves harnessing numerous less-intelligent devices, to vacuum a room, for example.

'The idea is that you try to give each robot a certain degree of independence, but also a little bit of co-ordination. If you get the balance right, you have the benefit of some co-ordination but a certain amount of autonomy that gives higher reliability,' he says.

'If one robot gets stuck or its battery runs out and the other is aware of that, it may say, 'Look, I'll clean over there'.'

Jarvis, who is the director of the university's Intelligent Robotics Research Centre, says another possible application involves patrolling large areas, where swarms bring an added benefit.

'A certain amount of randomness and uncertainty in patrolling is an advantage because whoever is trying to break in may not then be able to predict exactly when the robot will pass that particular door again.'

The inspiration, he says, has been social insects—mainly bees and ants—and researchers around the world are exploring similar approaches.

At the Universite Libre de Bruxelles in Belgium, for example, researchers built cockroach-sized (and scented) robots that successfully infiltrated—and, moreover, influenced—a community of cockroaches.

'What is new here is that the robot is autonomous. It is not remote-controlled by humans, and it acts at the social level in a group of living insects,' the project's leader, Dr Jose Halloy, told Reuters last month.

Similarly, the Defence Advanced Research Projects Agency in the United States has been testing bugs that can be controlled via computer chips injected into their bodies.

The agency has also been behind three challenges that give an insight into the pace of robotic developments.

In 2004, the agency offered $US1 million in prize money if an automated vehicle could navigate through 250 kilometres of the Mojave Desert in California. None of the robot cars lasted 12 kilometres.

The next year, with a $US2 million lure, five robot vehicles successfully sped through a tougher course of tunnels and hairpin bends, the fastest finishing in less than seven hours.

Last month the agency offered millions to attract robot cars to an urban course set on a disused air force base in Victorville, California, and it made the contest harder again.

At 96 kilometres, the course was shorter, but a six-hour time limit was set. The robot cars had to obey all traffic rules—including at a four-way intersection—merge with other traffic and avoid collisions. Six of 11 finalists succeeded.

Although he is not involved in the Defence Advanced Research Projects challenge (team leaders for the robot cars must be US citizens), Dr Thomas Braunl of the Robotics and Automation Laboratory at the University of Western Australia has been working on some of the techniques needed to make a car navigate safely without intervention.

These include brake-assist technology, which when alerted through laser sensors to impending danger can make non-invasive adjustments that nevertheless can help protect humans. Braunl, who has been working in robotics for 15 years, says these include increasing the tension of the seatbelt and reducing the gap before the car's brake pedal engages.

Elsewhere, the latest work in the field is traversing everything from underwater robots that can explore the harbour around Hobart and map it without intervention to flying robots that can detect—and avoid—aircraft flown by less-attentive humans.

None of these commercial ventures precludes scientists from one day conquering the age-old challenge of making a humanoid robot.

Durrant-Whyte compares the future of robotics with the history of computers, which have gone from mainframe machines that occupied their own room to BlackBerry-like devices that fit into a shirt pocket.

'These big [commercial] applications where you can afford to spend millions of dollars on a single robot, they're really the ones that are going to push it forward at the moment. They're like the old mainframes.'

But, as with the computer revolution, it will take years. In Japan, walking humanoid robots are already capable of simple tasks. And an Asimo robot handles basic reception tasks at Honda's Japanese headquarters, but there is still a long way to go.

In time, the advances funded by big commercial products will be incorporated in robots for domestic use and fulfil the early goals of eager scientists, complete with flawless skin, exquisite hands and a perky nose—all, no doubt, incorporating advances of their own to assemble the most beautiful of machines.

Ah, vanity, thy name is robot.

2

Faster, Stronger, Cheaper: Robotic Automation in the Workplace

Editor's Introduction

The assembly line, automobile magnate Henry Ford's great contribution to modern industry, is, among other things, a testament to the power of fragmentation. Under Ford's innovative system, large jobs, such as assembling an automobile, were broken into a series of much smaller ones, allowing workers to focus on a single task—riveting a car door to a chassis, say—and factories to maximize efficiency. Then as now, the trouble with the assembly line is that certain jobs can be dangerous and repetitive, and even if humans are able to meet performance standards, doing so often leaves them susceptible to injury and fatigue. It was for this reason that, in 1961, a General Motors (GM) plant in Ewing Township, New Jersey, installed the world's first assembly-line robot, a device that picked up red-hot car parts, such as just-cast molten-steel door handles, and dropped them into vats of cooling liquid.

By 2008, the number of robots being used at factories across the United States had risen to an estimated 178,000, with the worldwide total topping one million. While automakers have led the charge, companies in a variety of industries have come to realize the benefits of robotic automation. First and foremost, installing robots leads to a reduction in labor costs, since machines require only occasional maintenance and do not draw daily wages. What's more, robots can do things humans are unwilling or unable to do, such as grip hot metal or undertake the maddening task of inspecting thousands of McDonalds hamburger buns as they come rolling down a conveyer belt. It's perhaps for this reason that American workers have tended to embrace factory automation, despite fears robots may take jobs from human workers.

The articles in this chapter outline recent advances in robotic automation, examining how machines are helping everyone from car manufacturers to muffin makers increase efficiency. In the first selection, "Robots on the Rise," Jack Smith discusses "vision" and "force sensing," two of the areas in which robots are becoming more advanced. Today's machines have the ability to "see" products—a useful tool for inspection jobs—and gauge how much pressure to apply when gripping or assembling parts.

In "New Era of Robots," Lindsay Chappell interviews Gary Zywiol, vice president of product development at Michigan-based Fanuc Robots America, Inc. In addition to elaborating further on vision and force-sensing advances, Zywiol explains how the cost of implementing robot technology has decreased over the years, enabling companies to make use of cheaper, better machines. Touting robotics as an alternative to outsourcing, Zywiol tells Chappell, "And you know, people are moving offshore to reduce [labor] cost. We're trying to keep people onshore through automation."

In "Robotics Orders are Rising," the subsequent piece in this chapter, Michael LeGault calls 2007 "the year of the robot," citing as proof a 45-percent first-quarter increase in robot orders from automakers. Further quelling fears robots will replace human workers, efficiency expert Ron Harbour tells LeGault that such companies as Toyota and Honda have striven to maintain "a pretty good balance between man and machine." "It's proof you don't have to overautomate to be productive," he adds.

In "Robotics' Growth Skyrockets," Jean Thilmany focuses on how automation has transformed the baking industry. The fifth entry, "Machine-Tending Robots Increase Flexibility, Morale," looks at how employees feel about handing off duties to machines. "Our employees love the robots," Lloyd Grunvald, the manager of a Vermont mechanical-components company, is quoted as saying. "They have really elevated the employees from doing jobs that no one wanted to do in the first place to learning other more productive skills."

The final selection, "Meet the New Robots," finds writer John Teresko delving deeper into the subject of vision and explaining how some researchers are using robots to develop new cancer treatments.

Robots on the Rise

Showing Their Metal Mettle*

By Jack Smith
Plant Engineering, June 15, 2008

Since the first programmable robot design in 1954, much has happened but little has changed. While the first industrial robots were used in 1962 at a General Motors auto plant in New Jersey to spot-weld and extract die castings, nearly a half century later robots are still performing welding and material handling tasks.

When the robotics industry was relatively new, many people thought the intention was to replace workers with robots. To today's world-class manufacturers, automation is more about efficiency and flexibility than it is replacing workers. With manufacturers struggling to find skilled workers, robots are filling that void, as well as taking on tasks that can be performed safer or with less strain on the humans.

ROBOTIC MARKET TRENDS

Robots continue to be used in traditional applications, which include welding; assembly; machine loading, unloading and other material handling tasks; grinding, polishing and other material removal tasks; and palletizing. According to Jeff Burnstein, executive vice president of the Robotic Industries Association, the fundamental factors driving the use of robotics include:

- Lower costs for robot technology
- Improved reliability of robots and sensor technologies
- Faster, more flexible robots
- General industry needs for improved quality and productivity
- Industry specific needs for higher throughput, process validation and traceability.

Burnstein said as prices fall, new applications emerge in sectors such as drug discovery, food and beverage, consumer goods, plastics and rubber, furniture and fuel cells. For example, in the consumer goods industry, a major personal hygiene products manufacturer reported that robotic palletizing led to a 50% to 60% reduction in labor costs. In the furniture industry, a robotic packaging system increased production by 45% for a Swedish furniture producer. And in the food and beverage industries a major muffin maker uses robots to decrease damage to muffins resulting from mishandling after baking, according to Burnstein.

RIA estimates that some 178,000 robots are now at work in U.S. factories, placing the United States second only to Japan in overall robot use. More than 1 million robots are installed worldwide.

ERGONOMICS, MATERIAL HANDLING

In addition to welding, material handling has been one of the main reasons to invest in robotic automation. "There has been a steady increase in material handling applications, partially because of personnel injuries, health and medical claims," said Jay Sachania, director of marketing at Adept Technology Inc. "Plus, it's getting harder to find people to stand there working continuously doing the same thing over and over again."

The Chrysler plant in Belvidere, IL (a 2005 Top Plant winner) was retooled in late 2005 to accommodate the assembly of up to four different vehicles on one line. The body shop team took this opportunity to address an ergonomic issue.

Loading the front floor pan to an upright turntable was very difficult. The floor pans came from stamping stacked on top of each other in a horizontal rack. The operator had to bend over while reaching across the top of the pan, grab the other side of the pan and lift it, swing the front of the pan to a vertical position, carry the pan a short distance and place it in the turntable tooling. The team tried numerous assist devices with little success.

Using robotics, the team automated this task and resolved the serious ergonomic issue. The robot moves over the rack, lowers the tooling, senses the floor pan and uses suction-cup tooling to pick up one pan at a time out of the rack. It moves the pan to the conveyor and sets it in the conveyor tooling. The seat reinforcement is lifted into position under the pan. The robot picks up both parts and loads them onto the turntable, which indexes so another robot can weld the reinforcement to the pan.

Ergonomics also motivated SEW Eurodrive in Lyman, SC (a 2007 Top Plant winner) to automate its production. "One of the things that got us into automation was ergonomics—to avoid repetitive motion injuries," said plant manager Carl Hinze. "We took the burden off the employee."

At SEW, robots unload raw housings and move them to automatic milling stations where they are milled, bored, drilled and tapped. A typical workstation uses a main gantry arm and several stand-alone robots that load parts into and unload parts from the machines.

VISION

Using vision to guide industrial robots has increased robot-based assembly accuracy. Vision guided robots are now capable of simultaneous assembly and inspection. Locating—or even preventing—defects early in the assembly process greatly improves productivity by eliminating costly rework that could occur if defects are located later downstream.

"In high-speed packaging applications, vision is a vital component of the robot system as the location of objects coming down a conveyor line in a constant stream must be determined and tracked," said Ed Roney, manager of vision product development, FANUC Robotics. "These objects (often consumer products) flow to multiple robots, which in turn use the location information to make a successful pick-on-the-fly, and package the object into containers. Without vision, each of the parts would need to be well nested, which would be difficult given the high rate of manufacturing that is occurring."

Vision allows manufacturers to process various part types without robotic tooling changeovers. "Unlike blind robots, vision-guided robots don't depend on costly precision fixtures to hold parts, require additional labor to load and orient parts or need upstream actuators, sorters and feeders to separate parts for processing," said Bryan Boatner, product marketing manager for In-Sight vision sensors at Cognex.

In addition to providing manufacturing flexibility to accommodate product changes, Boatner said today's vision systems are a lot less expensive, which makes the cost of vision-guided robotic applications easier to justify. "Simplified calibration, easier integration and new connectivity standards make vision-guided robots faster and easier than ever to deploy. Beyond locating parts for pick-and-place or guiding a robot to assemble components, machine vision can also inspect, measure and read linear bar codes and data matrix codes as products are being handled or assembled," he said.

Most vision-guided robotic implementations use image analysis software to calculate positional information from a 2-D image, and provide it to a robot controller, according to Boatner. He said pick-and-place applications typically use a camera to acquire images of an area on a conveyor carrying objects for packaging, palletizing or assembly. "The vision system finds and computes the location of the objects on the conveyor, then converts the location into X-Y and theta coordinates, which it reports to the robot," he said.

Vision-guided robotic applications that fall somewhere between 2-D and 3-D typically use apparent changes in perspective or size to calculate 3-D data. "Layered bin picking presents objects in random positions and orientations in a stack of trays," explained Boatner. "When a camera views a stack of parts, the top object appears smaller as the stack gets shorter and its distance from the camera increases.

"In these types of applications the vision system uses the apparent change in size to calculate the top tray's height so that the robot can continue to add or remove parts from the stack. In the most complex applications, multiple cameras, or structured light techniques are used to provide 3-D data," Boatner said.

Part variability can hinder a vision sensor's ability to provide repeatable part location performance. Manufacturing processes have some inherent variability, which, according to Boatner, usually fall into these categories:

- Part rotation caused by lack of fixturing, vibration/motion on the line, etc.
- Changes in the scale of a part due to variations in the vision camera's optical settings
- Inconsistent or poor lighting
- Parts produced in distinctly different colors, textures, shapes and sizes
- Variations caused by modifications to the production process
- Substitutions of components or materials
- Different suppliers for a single part
- The presence of oil, paint, cleaning solvents and other substances that might obscure a part or change its appearance.

These may be accidentally introduced, or may be a known result of the production process.

If concerned about part variability, specifying vision software that supports sophisticated geometric pattern matching will ensure accurate and consistent part location, said Boatner.

Roney said the drivers of the robotic vision market include the flexibility to adapt to varying part presentations, lower costs from reduced tooling requirements and include online error-proofing requirements to verify that products are correct before further processing.

FORCE SENSING

Force sensing can help manufacturers eliminate damage from part insertion errors or improper alignment during assembly operations. The force with which a tool or end effector is used can be controlled by knowing the amount of force or pressure applied. There are several ways of accomplishing this, depending on how much manufacturers want to spend.

According to Thomas J. Petronis, vice chairman and CEO of Applied Robotics, many force sensing applications use servo current measurement. Although strain gauge and load cell applications exist, Petronis said servo current sensing is more cost-effective in manufacturing environments. Some medical applications or applications where high accuracy is required can justify the expense of these closed-loop systems. But for most industrial or manufacturing applications, servo current measurement is adequate and much less expensive.

Sachania said some robots use force sensing to test switches on automobile steering columns. Robots continuously actuate these switches during reliability testing. "These applications actually use force feedback to see how much force is being applied," he said.

However, Sachania suggests that force sensing is not widely used in manufacturing. Many manufacturing operations require speed of operation rather than accuracy of applied pressure. "You're looking for return on investment," Sachania said. "If the line is slow, your throughput is less. And if you want to be competitive in the market place, you want to make sure you are producing quality parts at higher speeds.

"For example," Sachania continued, "consider a packaging line that's picking and placing parts from one conveyor to the next, or picking up a part and putting it inside a box. If you try to add force sensing to it, you will slow down the line."

Sachania said that even with muffins, manufacturers are not using force sensing directly. Instead, systems that monitor the negative pressure applied to the vacuum cups or the torque applied to the motor provide feedback that offers force-sensing-type capabilities.

Up-front engineering and planning is critical. End effectors can be selected to apply several ranges of pressure. Based on testing during application design, parts can be picked, held and moved at high speeds without damage—even muffins.

END EFFECTORS

The end effector is end-of-arm-tooling, and is essential to the work that robots are required to do. Typical end effectors include spray guns, welding devices, grinders, deburring tools, vacuum cups and many varieties of grippers. Many end effectors are complex, and virtually all are application-specific.

Robotic tool changers greatly increase the versatility and flexibility of industrial robots. For example, at Chrysler's Belvidere Assembly Plant, robots equipped with special quick-change tooling end effectors position automobile body parts while welding robots join them. Robots can change their own tooling in less than one assembly cycle. Tool changers located adjacent to each robot station can hold several sets of tooling—one for each vehicle model that the Chrysler plant is now capable of running. Each robot is equipped with a standard end effector and tool mating assembly that attaches and releases very much like a socket wrench and ratchet.

THE FUTURE

Burnstein said the industrial robotics market, though maturing, still has a long way to go. "The nonautomotive markets such as life sciences, drug discovery, biomedical, plastics and rubber, and food and consumer goods hold great promise. Automotive opportunities will continue to grow as more cars are sold around the world. As the price of robots continue to fall, and their capabilities grow, look for new applications, such as robotic machining, to gain traction," he said.

New Era of Robots*

Seeing and Feeling

By Lindsay Chappell
Automotive News, August 9, 2006

The modern age of industrial robots can be divided into two eras.

In the 1980s, manufacturers realized the value of automating redundant and sometimes dangerous jobs, such as vehicle painting and spot welding. And in the past decade, advances in precision and dramatic reductions in the cost of materials, like chips and sensors, made new automation feasible.

In between, there was a period of re-evaluation, says Gary Zywiol, vice president of product development for Fanuc Robotics America Inc. in Rochester Hills, Mich.

The auto industry realized that some jobs simply didn't require automation. "There were some awful robot applications in the early '80s that gave robots a big black eye," Zywiol says. "And it was unwarranted. It wasn't the robots—it was that we picked the wrong applications in some cases. But we're over that hump now."

Zywiol talked about the business with Staff Reporter Lindsay Chappell.

What is it that customers are asking you for right now?

They want help reducing their overall costs.

Their labor costs, you mean.

They don't use the term "labor," but that's definitely on their mind. Automation doesn't have any direct labor cost in manufacturing products. There's obviously the maintenance factor, and you definitely have to increase the skills of your work force to be knowledgeable of the automation. But your direct labor cost goes down with automation. And you know, people are moving offshore to reduce that cost. We're trying to keep people onshore through automation.

How are you holding down the cost of the robot?

Robots cost less today in real dollars than they did 24 years ago when our company started. And at the same time, the performance is eight or 10 times what it was before. The cost of sensors is an area where we can reduce cost. We're seeking better and more affordable sensors. Just like in digital cameras, you can get higher megapixels at a lower cost than you could a few years ago. More megapixels means higher resolution. And in robotics, that means better accuracy.

Better vision allows robots to handle something else manufacturers want, and that's variation.

If you have a lot of variance of your product, it increases your production cost. So they're asking us for automation that can handle multiple products. We have many projects right now where we're using vision systems to identify products in multiple styles. They want to be able to follow the Dell computer model, where one line has the flexibility to produce different products in small lot sizes.

An application you might not have thought of is material handling. We have an application now where a customer is using vision technology on an overhead robotic arm that reaches into delivery trucks to unload a pallet. That would normally be performed by a person driving a forklift.

It sounds like vision technology is really driving the advances in robotics right now.

Ten years ago, maybe one-half to 1 percent of our robots had the ability to see. Right now, probably 5 to 10 percent have vision sensors.

That might not sound like a lot to you, but there are now more than 10,000 new robots going into U.S. plants every year. So that translates to maybe 1,000 robotic vision systems a year now.

It's definitely a trend for helping manufacturers do tougher applications, or jobs that are more human-intensive today that could be handled by robotics in the future. If I look 10 years into the future, I'd predict that about 50 percent of robots will have vision.

So yes, it's definitely a key technology. That and force sensing.

What's the application for force sensing?

Force sensing gives the robot the ability to feel. It will play a role in very difficult assemblies in the future, in powertrains and transmissions, for example, and the assembly of clutches, where there's a lot of intricate assembly work.

A robot today doesn't have the true touch-feel dexterity that a human does. Imagine putting two gears together and making those gears mesh. You could do that in your sleep. You can feel the teeth of the gears engaging.

The robot, on the other hand, can't feel. It's going to crush them together. Or in the assembly of a cylinder head—heads have either two or four valve stems to a cylinder, with incredibly small clearances to fit them into their holes.

You can't have a robot just stick those valves in. It's going to jam, it's going to miss.

Robots don't yet have the dexterity to feel whether gears have meshed correctly, or whether the valve stems are aligned correctly in an engine. But that's changing rapidly with advances in force sensing. You'll see more and more applications on jobs that once only humans could do. A ton of work has gone into the algorithms to provide sensor feedback to the robot, so that it can adapt its path as it works. You soften the pressure and allow it to react to forces to assemble tough pieces. There are some delicate jobs that could be handled by robots.

Is force sensing growing at the same rate as vision technology?

Not yet. But these are some pretty advanced applications we're talking about. Where the technology will head in market growth is not quite as predictable as the vision arena.

Vision sensing has really been following a predictable, almost chartable curve. Our own growth rate in vision systems at Fanuc has been a 30 to 40 percent increase in applications—per year—over the last five or six years. Even in years when the automotive market has been flat. That says something about the interest of the market.

With force sensing, people are still very cautious. Some of these applications are extremely difficult, even for us, the experts in the robot industry. We've got to make sure we spawn this business with successful applications.

How does the Internet play into this new era of robots?

The Internet enables networking. Robots have tremendously improved connectivity, and you'll see a lot more development in this area over the next 10 years. An example of this would be robots that can diagnose themselves. "I'm tired and I need a grease job," for example. They send an e-mail to the plant maintenance department or even to service people at our hot line at Fanuc Robotics. They say, "Hey, I'm a robot in Canton, Ohio, at such-and-such facility, and I'm predicting that in the next month I'm going to need preventive maintenance."

Another trend is what I call robot collaboration. This is multiple robots handling and assembling a part, with multiple arms working together.

Instead of one robot performing one job or two, we will see groups of robots working together, performing multiple functions. As we move deeper into this, you'll see less plant space consumed with conveyor space and fixtures.

The new connectivity is allowing this to happen. But there are roadblocks. Even though we have the Internet at work inside plants today, there is the issue of increased security risk. Factory information systems generally have lots of firewalls and protection to make sure that trade secrets are not let out of plants. But some of that technology is a bit stifled right now.

Are automotive customers pushing for applications that are simply not going to happen?

Some applications have related issues that have to be considered. One is safety —just from the point of view of plant layout. As you introduce more and more robots into factories, you start decreasing the distance between the human and the robot. Safety concerns have always been important, but you might reach the point where there's less room in a plant for enormous amounts of safety fencing, for example. So you will need to develop new standards and new solutions.

But I wouldn't say anything is a total dead end. It's a matter of deciding which direction we want to go. Final trim—yes, that's a very tough area to think about automating, theoretically. But maybe there are aspects of it that could be automated. Maybe the customer is thinking of automating some particular part of final trim, when we would instead offer an alternative of automating two steps up the line before you get to that operation.

Robotics Orders Are Rising

Suppliers Embrace Automation in Order to Compete*

By Michael LeGault
Automotive News, August 6, 2007

You won't find it on the Chinese calendar, but 2007 may go down as the year of the robot.

Robot orders from the automotive sector jumped 45 percent in the first quarter compared with the same period last year, according to the Robotics Industries Association.

"It's good to see orders up in the auto sector," says Jeff Burnstein, vice president of marketing and public relations for the Ann Arbor, Mich., trade group. "It's a pretty cyclical industry, so it's a sign the auto companies are gearing up new product launches."

Automakers and their suppliers ordered 3,238 robots in the first quarter, up from 2,233 in the same period last year.

Overall, new orders from North American robotics companies rose 24 percent in the first quarter from the same period of 2006. Most of the new orders were for material handling (36 percent of sales) and spot welding (31 percent of sales). Ake Lindqvist, vice president of global automotive sales for the automation giant ABB Group, attributes much of the sales boost to large orders from manufacturers.

"A few years ago, a typical (assembly plant's) body shop had about 300 robots," says Lindqvist. "Today there can be anywhere from 700 to 800 robots in a body shop."

Lindqvist says the large orders suggest automakers and large suppliers recognize that automation is the only way to compete in North America. He also says the robotics industry is taking the first-quarter sales spike in stride.

"The reason sales look so good in 2007 is because they were so bad in 2006," says Lindqvist, noting that 2005 was the first year robotic sales surpassed the industry peak in 1999.

"It's off to a good start, but let's look at the second and third quarters before we get excited," he says.

MAKING DECISIONS

Rick Schneider, CEO of Fanuc Robotics America Inc., says some automakers, especially the Detroit 3, had to cut costs in 2006. After the automakers decided which plants to keep open, they began spending to make their operations more efficient.

Auto companies have a strong incentive to buy robots, Schneider says. The machines perform some tasks better than employees—and at a lower cost, providing a better return on investment.

For example, improved vision capability gives robots an edge over workers in material handling.

"In the past, a person would have to manually orientate the part, which defeats the purpose." says Schneider. "Now a robot can go over to a bin of random parts, look into the bin, pick up a part and insert it into a machine."

Lindqvist says the Chrysler group is ahead of other North American automakers, including import brands, in integrating automation into factories.

Chrysler has "the most advanced flexible automation systems in North America," he says. "The transplant philosophy is to build many units of the same type, but Chrysler has realized it needs to be flexible and build different models on the same line."

Chrysler began increasing robotic investment several years ago as part of a plan to reach world-class efficiency and cost benchmarks at its operations. Achieving its goals requires a yearly 6 percent increase in productivity.

MAN VS. MACHINE

On the other hand, Asian automakers such as Toyota and Honda are reluctant to automate certain functions.

These companies use a lot of robots—and will use more of them. But they also understand that people won't break down like machines and won't need to be rebooted.

Toyota maintains "a pretty good balance between man and machine," efficiency expert Ron Harbour told *Automotive News* in an interview last year. "It's proof you don't have to overautomate to be productive."

But Chrysler spokeswoman Michelle Tinson says the key to Chrysler's automation strategy is a three-pronged approach in which robots perform:

1. All tasks in assembly plant body shops.
2. As much material handling in plants as possible.
3. A variety of manufacturing duties.

"We see our flexible robotics initiative as a prime driver of cost reduction and productivity improvements," says Tinson.

She says flexible manufacturing has been successful at a number of Chrysler sites—such as the Belvidere, Ill., assembly plant, which builds the Jeep Compass, Jeep Patriot and Dodge Caliber.

The new Chrysler Jeep plant in Toledo, Ohio, and its adjacent Supplier Park are loaded with the latest generation of robots.

Tinson says Chrysler's aggressive automation strategy is fueled by improved robotic technology, which makes the machines more useful, and by declining robot prices.

Robot makers say sales growth in 2007 and beyond will depend not just on automakers but also on suppliers, whose investment in robots has declined. ABB's Lindqvist doesn't think cost is the main issue at many suppliers.

"There's a lack of integration capacity at the Tier 2 and Tier 3 levels," he says. "System integration is a risky business, and there are not a lot of great integrators out there."

He says it's a problem the industry is trying to address because "the auto business drives the food chain."

Robotics' Growth Skyrockets*

By Jean Thilmany
Baking Management, October 1, 2008

While robots are no strangers to the wholesale baking industry, explosive growth has occurred during the last five years as their costs continue to fall and the number of jobs they can perform grows. Robots and their attendant hardware and software—the robotics that can automate so much of a baking facility—are now gaining a definitive foothold with wholesale bakers.

Benda Manufacturing Inc., Tinley Park, Ill., formerly responded to only one or two queries about robotics applications for wholesale bakers each year. Now, potential customers call weekly asking how robotics can help automate their operations, says Terry Benda, president of the food-handling systems company. "Robotics are booming," he says. "Within the last five years, I've started seeing them in larger roles where they're replacing operators."

Many more wholesale bakers are considering robotics implementation because the automation technology's costs have dropped by at least 50 percent in the past decade, says Rick Hoskins, vice president of operations, Colborne Corp., Lake Forest, Ill. Colborne produces automation equipment for the baking industry.

Why the steep price slash? Simple economies of scale, says Charles Gale manager, automation sales, Weldon Solutions, York, Pa., a systems integrator and robotics provider. "More people are buying them, so we can make them cheaper," Gale says.

Today's robotics applications also are easier to maintain than in years past as robotics makers take advantage of ever-advancing technological innovation, Gale says. Advances also mean robots now perform faster than older models and can keep up with high line speeds.

These new robots can offer cost reductions to bakers in a number of ways. "In almost all cases, the large portion of value is in labor reduction," Hoskins says. "However, in many cases there are other automation or mechanical machines available to automate the same basic tasks. In these cases, the value of robotics is typically reliability because you have fewer moving parts and more flexibility, and because robots are easily programmed, which makes it easier to adapt to different products or characteristics," he adds. "The changeover is much faster as well."

NEW ON THE LINE

Northeast Foods, Baltimore, turned to robotics four years ago for line balancing and found significant cost savings via labor reductions and reduced machine-operation costs. The bakery, which supplies buns to more than 500 McDonald's restaurants, was no stranger to automation. Much of its plant operations had already been automated before executives turned their attention to the conveyor. Prior to introducing robotics, two employees were responsible for final inspection and line balancing.

Though the employees did their jobs with accuracy and precision, Northeast Foods decision makers realized the repetitive job and the challenges of inspecting and balancing 5,000 dozen buns per hour meant new employees wouldn't exactly be keen on steping into the role when the present inspectors moved on.

They turned to two FlexPicker robots from industrial robot maker ABB Robotics, New Berlin, Wis. Northeast's new robotics system includes a vision camera trained on all six lines in the plant. Software tied to the camera sends detailed information to the robots about gaps coming down the line. The software instructs the robot to move buns from full lanes to empty lanes to achieve a balanced flow of product.

"Since we began running the line in July 2004, equipment downtime has been minimal. The robot requires very little maintenance aside from replacing the rubber suction cups that pick up the buns," says a Northeast Foods' representative. The suction cups need to be replaced every three to four days because of the high-speed production.

The robots do their work as the conveyor moves 120-ft.-per-minute. Northeast Foods worked with ABB to ensure the robots could meet Northeast Foods' 80-parts-per-minute handling requirement.

Collaboration between baker and robotics supplier leads to another reason for robotics' growing popularity, Gale says. Robots today can be easily programmed to perform a variety of specialized applications and to handle products in a particular manner.

For instance, the robots that package the éclairs made by the Italian pastry maker Forno Bonomi, Rovere Veronise, Italy, use a decidedly light touch when picking up the pastries. Four robots—also from ABB Robotics and tied to a vision system—pick the éclairs from the belt and gently place them on plastic trays in three layers of seven. The robots display that delicacy thanks to the gentle grip of the suction cups, which Bonomi worked with ABB to design, says Renato Bonomi, who co-owns the baking operation with his brother. The robots also are flexible enough to work in a narrow space, he adds.

But bakers don't need to necessarily team with a vendor or integrator to program their robots. Now, robotics suppliers, such as Weldon Solutions, can train a customer, who has never even seen a robot up close, to write a software program that will automatically guide the robot to perform a set of tasks. Training takes about 15 minutes, Gale says. "Before, you'd have to hire someone to do this for you. But robotics companies are making it much easier to rewrite programs and making them user-friendly. Should a robot not function as desired, or should a new function be sought, the robot can usually be reprogrammed in an afternoon," he adds.

MORE AND VARIED WORK

Robotics also has seen its popularity surge with bakers because of the stepped-up list of tasks these advanced machines can perform on the baking line. The robots of 10 years ago moved in only two dimensions, forward and back or up and down, which meant they could only carry out limited operations, such as physically picking up and moving an object a short distance, Benda says.

Today's robots demonstrate a much wider range of motion, and they're often tied to vision systems, which essentially gives them sight and allows them to carry out a number of tasks not possible even a few years ago, he adds.

These physically versatile robots have a stepped-up role to play in the wholesale baking industry. They now are automatically packaging products at the end of the line, and picking and packing orders for shipping. When tied to a vision system, robots balance production by reading line gaps to ensure conveyors are properly filled with product, as at Northeast Foods.

"Inline pan balancing makes the system more efficient and reduces injuries," Benda says. "It's a physically demanding job to handle hot and heavy pans all day long." Robots also are regularly moving into roles formerly handled by manual operators, such as forming patterns and then loading the product into trays and cartons. Today's robots also are capable of stacking and unstacking baking pans, trays and peel boards, Benda adds.

Order picking is another new job for robots. "The bakery route driver is making up his orders for the day and has to pick product from multiple stacks of trays to build a complete order for a customer. These stacks are typically 8 ft. tall, making for a very difficult and time-consuming task," Benda explains. "Robots can now automatically build these orders." In some cases, they can even load the truck.

Today's robots also have a role to play in the swiftly changing package sizes now demanded by consumers and retailers, says Ray Anater, director of automation sales and development at automation supplier LeMatic Inc., Jackson, Mich. His company has been following a trend of late that calls for baked products to be packaged in ever-changing quantities. Robots can scale to package these varied amounts, he adds.

"Traditionally in the baking industry if you make hamburger or hotdog rolls, they would be packaged in an eight-count bag or a 12-count bag," Anater says. "But now, we are seeing a move toward smaller packaging sizes. So, that means the packaging equipment also needs to be more flexible, to scale up or scale down. And now the robotic handling of the equipment is inherently flexible, which will drive the use of robotics in packaging."

LeMatic also produces robots tied to a vision system, capable of balancing products on the conveyor line and loading English muffins, buns and croissants onto trays. "The traditional method of using gravity slides to guide products isn't 100 percent effective," Anater says. "Either they get stuck or hung up, so an operator is needed to balance products across the lanes."

Like other robotics providers, LeMatic expects to see sales grow as bakers discover robots increased ease of use and as they consider how robotics can give them a competitive edge.

Any residual doubts should fade away as wholesale bakers speak with each other and meet with system integrators to discuss needs, Gale says. "Clients have a higher comfort level than they did because they hear these success stories, and they're trying to find ways to incorporate robotics into their operations," he adds.

So, what of the future? More of the same, only better, Benda says. Robots will become an everyday part of the operation, performing line balancing, tray stacking and order picking on a regular basis. Both Hoskins and Benda look forward to robots that can perform cleaning duties in a plant's wash-down area.

They agree that when it comes to wholesale baking applications, robots are here to stay. "More and more, bakeries will find out that it's economical to use robotics," Benda says. "That's really the future."

Machine-Tending Robots Increase Flexibility, Morale*

Modern Applications News, September 2004

Preci-Manufacturing, Inc. (Winooski, VT) is a family-owned manufacturer of precision mechanical components for the aerospace, defense, medical, and electronics industries, which relies heavily on machine-tending robots as a vital and productive part of its close tolerance machining process.

Located in the foothills of Vermont's Green Mountains, Preci-Manufacturing's 40,000 sq.-ft. facility employs three Rixan/Mitsubishi 6-axis robots—one that has been running almost continuously for six-years and a second for almost five-years.

A third 6-axis robot is newer and working daily, according to Lloyd Grunvald, general manager, who runs the company with brother Jeff and father Marcel. The "Preci" name was derived from the first five letters of the word "precise." A company specializing in precision machining and holding tolerances of up to +/- 50 millionths would expect nothing less in the machine-tending robots they employ.

From his experience with the first two robots, productivity is also expected. "They have been running for five-years with no down time," explains Grunvald.

"Having the payload capacity and reach we sought, as well as the Mitsubishi reputation for quality, consistency, and accuracy, the Rixan/Mitsubishi 6-axis robots were perfect for our needs in automating our machining processes. We have not been disappointed as we get 100% efficiency and maximum flexibility with our Rixan/Mitsubishi robots. They are extremely dependable," Grunvald commented.

JOB LOT FLEXIBILITY

"Prior to adding the robots," says Grunvald, "our machine operators were spending more time loading and unloading machines than they were in producing and inspecting finished parts. Now, we can run jobs that vary from 200 to 50,000 pieces—one operator runs three machines—and our productivity and throughput have increased dramatically. The robots allow for quick-change, end-of-arm tooling so we can switch between machining jobs with very little downtime or difficulty.

"We have also developed a piece of equipment," notes Grunvald, explaining that it is, "rather like a horizontal slide which is affixed to the CNC machine. It allows for the robot to be moved easily from side-to-side to afford operator access to a machine, if necessary. However, the heart of the slide is the attached 'parts holder platform,' which can be adjusted to the pre-taught positions of the robot.

"This saves us an enormous amount of time," advises Grunvald, "because we don't have to 're-teach' the robot when we change a job. All we do is adjust the part-holder platform to the robot."

Grunvald adds, "We are also working on a portable system where we have a robot affixed to a table which is wheeled right up to a CNC machine, hooked up in a matter of minutes, and ready to start loading/unloading parts. It is still in its developmental stage but we are very excited about its possibilities.

"This flexibility is giving us the opportunity to handle larger volume, difficult-to-machine materials. One of the new areas we have started working in and exploring further is the automotive market," Grunvald comments. "We see huge opportunities for Preci in this area because of the precision we offer. Of course, robotics and automation will play an even more crucial role in meeting the close tolerance demands and 'Just-In-Time' (J-I-T) delivery of parts for the automotive industry."

EMPOWERING EMPLOYEES

Asked about employee reaction to the introduction of robots, Grunvald notes, "Our employees love the robots because they have taken away the previously-dreaded mundane loading/unloading tasks. They have really elevated the employees from doing jobs that no one wanted to do in the first place to learning other more productive skills, like programming and maintaining the robots and inspection of machined/finished parts."

Integration of the robots into the process was simplified by free employee training provided by Rixan/Mitsubishi.

"We have had no lay-offs as a result of purchasing and using the robots," the general manager states. "We have had a tremendous upsurge in productivity and employee esteem, and just as important, an increase in sales. The robots allow us to more efficiently handle small jobs, large jobs, and J-I-T delivery jobs.

"We feel that by automating and utilizing robots, it has helped us overall in growing and competing in the numerous markets we serve as well as opening the door to new markets.

Commenting on the state of the manufacturing industry in the United States, Grunvald adds, "rather than 'off-shoring' and losing jobs and manufacturing businesses, American manufacturers need to work smarter, faster, and better. Productivity levels can be vastly increased by incorporating automation.

"In order for companies to compete and stay alive, now, and in the future, there has to be a commitment to investing in capital equipment. We started five years ago with robotics and automation and what a dramatic difference it has made," observes the general manager.

Meet the New Robots*

By John Teresko
Industry Week, October 1, 2007

The time: 1961. General Motor's implementation of flexible robotic automation in Ternstedt, N.J., started U.S. manufacturing on the path to realizing a future depicted in a 1923 play by Karel Capek. In "R.U.R." (Rossum's Universal Robot), Capek's vision was for millions of mechanical workers—robots (as derived from the Czech words for work or workers).

Although U.S. robot numbers are not yet measured in millions, the industrial automatons are nonetheless playing strategic roles in U.S. manufacturing competitiveness, says Jeffrey A. Burnstein, executive vice president, Robotic Industries Association. RIA estimates that more than 171,000 robots are now at work in U.S. factories, placing the U.S. second only to Japan in overall robot use. Worldwide, there are more than a million industrial robots in operation, Burnstein notes.

To properly appreciate the value contribution of that installed base, consider that those mechanical workers are for the most part succeeding despite being blind, deaf and without a sense of touch.

But that's changing. For example, robotic vision and other intelligence features were strong trends at the recent RIA 2007 International Robots & Vision Show. One example is a new intelligent welding robot from Fanuc Robotics America Inc.

"In a welding cell, a multi-arm control allows customers to achieve maximum utilization of their robots," says Mike Sharpe, Fanuc Robotics' director of engineering for materials joining. "Using one or two robots for material handling while other robots perform welding maximizes flexibility. By simply changing the end-of-arm tooling on the material handling robots, a wide variety of parts can be welded within a single work cell." An Ethernet-based welding network allows the controller to handle up to four welding power supplies. Sharpe says worldwide installations of Fanuc robots number more than 172,000, with over 80,000 installed in North and South America.

Motoman Inc. demonstrated a new vision-guided robot solution at the robotics show that could pick randomly located automotive components out of a bin and place them on a table. The robot then individually placed the parts into another bin. The Motoman robot uses the vision system to access part position information using either serial or DeviceNet interfaces. The vision software supports true 3D (X, Y, Z, yaw, pitch, roll) with one, two or three cameras, without the use of range sensors or lasers. The single camera option requires a robot mounted camera and multiple inspections. With the multiple camera option, the cameras can be mounted on the robot or at a fixed location. That allows greater flexibility and can accommodate irregular part shapes. The technology makes complex bin-picking applications possible, despite confusing backgrounds.

Motoman says the customizable human machine interface (including menus) allows users to change system parameters, calibrate the system and see real-time inspection results. Integrated 2D solutions are also available.

A variety of pressures are accelerating the convergence of vision solutions with robots. Contributing to the process is the recognition by manufacturers that many robot applications lacked the flexibility to easily accommodate the smaller batches and frequent changeovers required for mixed-model processing, says Cognex Corp.'s Bryan Boatner, product marketing manager for In-Sight vision sensors. "Unlike blind robots, vision-guided robots (VGRs) don't depend on costly precision fixtures to hold parts, require additional labor to load and order parts or need upstream actuators, sorters and feeders to separate parts for processing. Consequently, VGRs allow manufacturers to more easily process various part types without tooling changeover. Plus, VGRs provide the added benefit of automatic collision avoidance for safer work cells."

Adds Greg Garmann, software and controls technology leader for Motoman, "It can be a trade off. By adding more intelligence to the robot and camera systems, the hardware and tooling investment can be simplified and diminished."

In addition to providing manufacturing flexibility to readily accommodate product changes, Boatner says today's vision systems are a lot less expensive. For example, in the mid-1980s a flexible manufacturing system went online sporting an elaborate $900,000 3D robot guidance system. By 1998, the average cost of an implementation was down to $44,000, according to the Automated Imaging Association. And the downward spiral didn't stop there.

"Vision has never been more affordable [than now]," says Fanuc Robotics' Dick Johnson, general manager, material handling. For example, Fanuc Robotics offers 2D visual robot guidance for $7,995 and visual error proofing for $4,995 with its robots.

"In robotics, the important factor contributing to price reductions was the emergence of the automakers as the large volume early adopters of the technology," says Kevin Kozuszek, director of marketing, Kuka Robotics Corp.

OFFERING SIGHT TO THE BLIND ROBOTS

Vision hardware is becoming much more reliable, notes Johnson. "At one point vision algorithms ran on expensive, complex dedicated hardware. This was then replaced with systems that made use of personal computer hardware." He reports that the new trend is to offer vision built right into the robot, as with the Fanuc Robotics iRVision, or supply small cameras from companies like Cognex. "The elimination of hard disks and operating systems designed for personal use greatly increase the system reliability."

Johnson also notes that programming and calibration have become easier and less time consuming. "This means that the robotic vision system comes up and runs faster." Johnson reports that robot suppliers are also making it easier to interface the results of the vision system to the robot.

Expect vision systems with simpler lighting requirements, says Johnson. "Vision systems are becoming more immune to lighting variations. In the case of Fanuc Robotics vision, the programmer can take advantage of multi-exposure control to snap the same image with different exposures. That allows the vision algorithm a wider range of operation and the ability to compensate automatically for lighting variations through the day." Optional ring lights also simplify the engineer's job of properly lighting the part. The ring lights attach directly to the camera and provide a ring of LEDs around the camera lens. This ensures that the light is directed right where the camera needs it, explains Johnson.

The need for visual error proofing is also driving the robot/vision convergence. Early robots were sightless, leaving manufacturers to contend with the loss of operator feedback on process abnormalities. Johnson's example: "An operator might note that a part is missing a feature or that a label is being placed upside-down. Blind robots won't be aware." While low-cost error-proofing systems need to be programmed, they will perform even better than the human operator, Johnson says.

"As manufacturers strive for increased quality, the vision solution can offer an advantage," Johnson continues. "Six Sigma quality systems, for example, allow only 3.4 defects per million parts. While an operator may tirelessly check 10,000 or even 100,000 parts for a given defect, a properly programmed robot will find all three defects in a run of 1 million parts. The trend toward use of visual error-proofing allows robotic systems to improve quality by checking and taking action to reject defective parts, thus saving on the costs associated with scrap, rework, repair and warranty."

Vision capability and accompanying accuracy improvements are helping to spur more diverse robotic applications. Two examples were honored at the robotics show as winners in the RIA's user recognition program. The High Throughput Screening Core facility at the Memorial Sloan Kettering Cancer Center in New York, for instance, uses robots and machine vision to screen large chemical libraries against various cancer targets. "For economical reasons, many pharmaceutical companies aren't willing to research certain types of rare cancers," explains Dr. Hakim Djaballah. The system was designed internally at the center and integrated by the laboratory automation and integration group at Thermo Fisher Scientific.

Molding International & Engineering uses a vision-guided robot system to manufacture insert molded electrical connectors. Using end-of-arm tooling, the claimed benefits include increased plant capacity, quality improvements, reduced variation in molding processes and elimination of work-in-progress inventory. Automation suppliers include Denso Robotics and Tensor Automation.

3

"The Robot Will See You Now": Robotics in Health Care

Editor's Introduction

The benefits of building smarter, more versatile robots aren't only being realized on assembly lines. In hospitals and health-care centers around the world, machines are running supplies to nurses, sorting medication, and even performing surgeries—their precise cuts and ability to grasp tiny instruments leading to smaller incisions and shorter periods of recovery. These are but three of the ways robots have already made their mark on medicine, and according to researchers, still-greater breakthroughs are just around the corner.

One example is HAL, or "hybrid assistive limb," the robotic exoskeleton writer John Boyd discusses in "Dress for Action with Bionic Suit," the first article in this chapter. Created by Japanese inventor Yoshiyuki Sankai, the HAL suit might one day restore mobility to the elderly and paralyzed and allow able-bodied individuals to lift objects far heavier than they could heft on their own. Amazingly, the suit's motion is triggered by signals sent from the brain, and it responds quicker than would the wearer's actual muscles.

In the next selection, "Robo-School," Jean DerGurahian profiles the International College of Robotic Surgery, a 410-bed learning facility run by St. Joseph's Hospital in Atlanta. The school teaches surgeons how to use the da Vinci Surgical System, a robotic arm capable of performing a variety of operations. While some surgeons have been reluctant to adopt the new technology, which requires costly, time-consuming training, the college's lead instructor, Sudhir Srivastava, tells DerGurahian, "I think surgeons will be almost forced to learn this." Robot-assisted surgeries are generally less invasive than traditional procedures, and the resultant shorter hospital stays translate to lower costs for hospitals and patients, making them an attractive option for all involved parties.

The following two pieces deal with microbots and nanobots, two types of minuscule machines that, while still largely theoretical, have captured the public's imagination. In "Now You See Me . . . " Claire O'Connell introduces the concept of swarming microbots, or mechanized critters that scientists believe could work in tandem and carry out medical procedures inside patients' bodies. Microbots, roughly the width of a human hair, are far larger than nanobots, which would measure a mere billionth of a meter. Nanobots are more likely to be built from chemicals than traditional robotic materials, O'Connell explains, likening them to "souped-up molecules."

"Swallow the Surgeon," the subsequent entry, opens with a reference to *Fantastic Voyage*, the famous sci-fi story in which doctors shrink themselves to the size of microbes and search for a blood clot inside the body of a dying patient. While such a procedure will likely never be feasible, the article's author describes efforts to create robots small enough to enter the eye, stomach, or digestive tract. The story focuses on the challenges associated with guiding such robots through the body.

In "Robots: The Next Generation," Joseph Mantone discusses robots designed to increase efficiency at health-care facilities by eliminating the need for people to carry out mundane tasks. Mantone looks at robot couriers and pill sorters, as well as a machine that enables doctors to make their rounds from remote locations. The doctor appears on the robot's video-screen head and uses a remote-control camera to examine patients. "There's a chuckle factor involved," Charles Casey, a spokesman for the University of California-Davis Medical Center tells Mantone. "On its surface, it's strange, but everyone has liked it."

In the final piece, "Paging Dr. C-3PO," Dave Carpenter introduces the Safe Accurate Medications for Inpatients, or SAMI, a "giant jukebox" of a robot that doles out prescription drugs faster and more accurately than human pharmacists. "What used to take pharmacists about six hours—an overnight fill of the next day's prescription—with the chance that there could be errors made, now takes an hour and a half and it's flawless," Dr. Donald Manning, chief medical officer for the company behind SAMI, tells Carpenter.

Dress for Action with Bionic Suit*

A Motor-Driven Exoskeleton Will Give Its Wearer Freedom of Movement and Super-Strength

By John Boyd
New Scientist, April 9, 2005

A robot suit has been developed that could help older people or those with disabilities to walk or lift heavy objects.

Dubbed HAL, or hybrid assistive limb, the latest versions of the suit will be unveiled this June at the 2005 World Expo in Aichi, Japan, which opened last month. A commercial product is slated for release by the end of the year.

HAL is the result of 10 years' work by Yoshiyuki Sankai of the University of Tsukuba in Japan, and integrates mechanics, electronics, bionics and robotics in a new field known as cybernics. The most fully developed prototype, HAL 3, is a motor-driven metal "exoskeleton" that you strap onto your legs to power-assist leg movements. A backpack holds a computer with a wireless network connection, and the batteries are on a belt.

Two control systems interact to help the wearer stand, walk and climb stairs. A "bio-cybernic" system uses bioelectric sensors attached to the skin on the legs to monitor signals transmitted from the brain to the muscles. It can do this because when someone intends to stand or walk, the nerve signal to the muscles generates a detectable electric current on the skin's surface. These currents are picked up by the sensors and sent to the computer, which translates the nerve signals into signals of its own for controlling electric motors at the hips and knees of the exoskeleton. It takes a fraction of a second for the motors to respond accordingly, and in fact they respond fractionally faster to the original signal from the brain than the wearer's muscles do.

While the bio-cybernic system moves individual elements of the exoskeleton, a second system provides autonomous robotic control of the motors to coordinate these movements and make a task easier overall, helping someone to walk, for instance. The system activates itself automatically once the user starts to move. The first time they walk, its sensors record posture and pattern of motion, and this information is stored in an onboard database for later use. When the user walks again, sensors alert the computer, which recognises the movement and regenerates the stored pattern to provide power-assisted movement. The actions of both systems can be calibrated according to a particular user's needs, for instance to give extra assistance to a weaker limb.

The HAL 4 and HAL 5 prototypes, which will also be demonstrated at Expo 2005, don't just help a person to walk. They have an upper part to assist the arms, and will help a person lift up to 40 kilograms more than they can manage unaided. The new HALs will also eliminate the need for a backpack. Instead, the computer and wireless connection have been shrunk to fit in a pouch attached to the suit's belt. HAL 5 also has smaller motor housings, making the suit much less bulky around the hips and knees.

HAL 3 weighs 22 kilograms, but the help it gives the user is more than enough to compensate for this. "It's like riding on a robot, rather than wearing one," says Sankai. He adds that HAL 4 will weigh 17 kilograms, and he hopes HAL 5 may be lighter still.

Sankai has had many requests for the devices from people with brain and spinal injuries, so he is planning to extend the suit's applications to include medical rehabilitation. The first commercial suits are likely to cost between 1.5 and 2 million yen ($14,000 to $19,000).

Robo-School*

By Jean DerGurahian
Modern Healthcare, July 6, 2009

The doctor makes a movement and the knife slices fluidly through the muscle, parting the tissue as easily and cleanly as scissors cut silk. The movements go on like this, and the surgical tool is wielded confidently, efficiently, never shaking or deterring from the path it is supposed to take as the procedure continues on video.

Except that the surgeon is sitting in the operating room a few feet away from the patient, and the movements are generated through the video console. It's the surgical arm, the robot, that does the actual cutting and remains oblivious to any potential mistakes that would otherwise run through a human mind.

The robot is the da Vinci Surgical System, manufactured by Intuitive Surgical, Sunnyvale, Calif., essentially the only such robotic arm on the market. It's not new—the da Vinci system has been around for the past decade and has been installed in more than 850 hospitals—but despite claims by many doctors that the robot makes minimally invasive surgeries easier to perform, providers aren't clamoring to use them.

Part of the reluctance is on the part of surgeons who aren't sure about learning to use a machine when they can perform the procedures well using their own hands. St. Joseph's Hospital of Atlanta is stepping in to help smooth out what might be considered a daunting learning curve for some.

BACK TO SCHOOL

The 410-bed hospital launched the International College of Robotic Surgery earlier this year as part of its effort to meet a demand for training. While Intuitive has previously established centers—typically done in partnership with various hospitals, including St. Joseph's, to train physicians on the surgical system—St. Joseph's college wants to take that basic training to the next level.

The college—which is starting with cardiac surgery but plans to expand into other types of procedures—hopes to train surgical teams on advanced, minimally invasive robotic techniques using customized education modules based on various levels of expertise and need. The curriculum includes Internet course work, quizzes and cases to complete in order to "graduate."

The college offers interactive, online and remote guidance to teams from around the world interested in learning about robotics. Medical teams are invited to travel to Atlanta and attend sessions in person as well.

St. Joseph's experts also will visit a team's site to provide follow-up support and training.

Costs for the training can range anywhere from $4,000 for online courses up to $100,000 for the full, one-year curriculum, based on how many people are in the surgical team, whether they want to visit in person or take virtual lessons.

Using a combination of hands-on, virtual classes and videoconferencing, the doctors at St. Joseph's who lead the college hope to be what they themselves didn't have: mentors.

It's difficult not to sense Sudhir Srivastava's excitement about the college or commitment to robot-assisted, minimally invasive surgery. The cardiothoracic surgeon was brought onboard in February by St. Joseph's to be one of two surgeons who lead the college and conduct training. Srivastava serves as president and chief scientific officer for cardiac revascularization. His counterpart, Douglas Murphy, also a cardiothoracic surgeon, serves as chief scientific officer for intracardiac robotic surgery for the college and chief of cardiothoracic surgery at St. Joseph's.

Several components are involved when learning how to use the robot, says Srivastava, who has performed more than 1,000 robotic cardiac surgeries since 2002. About a third of those were conducted on beating hearts, a procedure that demands a level of skill that can only come with significant practice, he says. By using the latest technologies and techniques, that's the level he and Murphy hope to bring to others faster through the college.

"We went through a lot of learning curves," Srivastava says. "People don't have to go through that extended learning curve or frustration."

The college is approaching robotics training from the team standpoint because robotic surgery "is truly a team effort," Srivastava says. There is the surgeon at the console, watching the video and manipulating the controls that tell the robot's arm how to move. And there are the nurses at the patient bedside, monitoring the patient and performing tasks to keep the surgical site clear for the robot.

The training is as much about learning how to work well as a team as it is about learning how to manipulate electronic controls, he says. The surgeons are learning hand motions via remote visual cues through a computer screen, but everyone also is learning how their team members function and communicate, especially if there are any disruptions or problems that arise during surgery. "In three months or so we can get people comfortable."

Getting comfortable is just the beginning, however. After learning to control the machine and improving team communications through simulations, the college recommends doing as many cases with patients as possible to understand how the machine and patient interact—it's much like going through clinical training all over again, Srivastava says. He and Murphy will follow up with teams that have come through the college in their facilities in the first year. "We want to see what they've accomplished," he says.

Since opening in January, the college has had about four to five teams come through for training, both virtually and in person, each month. Some have come from as far away as Shanghai to spend a week. The college is working to update its Web site with more content and training modules. Srivastava says he would like to eventually see the college training eight to 10 groups each month.

Srivastava is certain that robotics surgery will grow, as more patients demand the procedures that get them in and out of the hospital more quickly, and as providers see the benefits of having such machines.

"I think surgeons will be almost forced to learn this," Srivastava says. "The technology is continuing to get better and be more surgeon-friendly."

A GROWING MARKET

According to market research conducted by BCC Research, Wellesley, Mass., use of medical robots and computer-assisted surgical equipment is growing in the U.S. The market was worth about $648 million in 2008 and is projected to reach $676 million in 2009, up 4.3%. By 2014, the market will be worth $1.5 billion, according to the report, up about 130% from 2008. Surgical robot systems were the largest product in that market as well, accounting for 54% of the market share. That share is expected to increase to 65% by 2014, according to the BCC report.

With that growth occurring, the design of training remains critical. And it's an issue that has been a concern among physicians for some time. There are a range of surgical opportunities that open up through the use of robotics, writes Richard Satava, a physician who is a professor of surgery at the University of Washington, Seattle, in a 2007 article. His column was published in the *Bulletin of the American College of Surgeons*, where he discussed the need for the right kind of training.

The robotics console "is the overall architecture that will provide even greater capabilities in the future—and this is just the beginning," Satava writes. But without first incorporating education that matches high-tech surgery, robotics can't reach its full potential. "Thus, it is necessary to incorporate the basic principles of adult education, curriculum design, setting of quantitative performance metrics for outcomes and validation of the curriculum," he writes.

Murphy, with the robotics college, agrees. "Robotics surgery is still in the pioneer stage," he says. There is a deep body of knowledge about robotics that hasn't been available to all physicians, and there wasn't a way to be trained. He views his role and his colleague Srivastava's role at the college as being one of leaving a legacy. "We wanted to put some of our effort into teaching other people rather than just do a bunch of cases," he says.

The college Web site has a full curriculum, including its da Vinci Connect program to broadcast presentations into hospital conference rooms; practice laboratories; and the teaching of small procedures that focus on gaining proficiency with the technology, he says.

Beyond just training, there are other obstacles to adding robotics to surgical programs, Murphy says. The hospitals have to be willing to pay more than $1 million for the system, plus maintenance fees, and surgeons need time to access the systems. Also, while doctors are learning how to use the machines, their productivity drops, Murphy says. "It takes real institutional commitment to have a robotics team."

Still, it's worth it, Murphy says—a point he hopes to help drive home through the robotics college. Being proficient at using the robotic system has enhanced doctors' technical abilities and has allowed them to perform safer surgeries on patients who didn't have the minimally invasive option before, he says.

Patients who are obese, who have thicker muscular structures and broader chests, or patients with abnormal skeletal structures would have required open surgeries, because doing minimally invasive surgeries on them is too difficult, Murphy says. With the robotics involved, doctors can gain access more easily.

Financial incentives also play a role. "You don't get paid more for doing the surgery robotically," Murphy says. At the same time, a surgery using the robotic arm can cost $12,000 to $15,000 less than a typical surgery, with fewer complications and a shorter length of stay, he says.

Those metrics alone should make providers consider robotic surgery with more enthusiasm, says Julian Schink, chief of gynecologic oncology at 787-bed Northwestern Memorial Hospital in Chicago, which has had the da Vinci equipment since 2007. Schink has not undergone robotics training through St. Joseph's college, but has been trained through the da Vinci system's manufacturer, Intuitive.

Patients experience less bleeding, less pain and are back to work faster after procedures done robotically, which saves costs for hospitals, Schink says. Once his team demonstrated that the number of hospital days among oncology patients was down 60% to 65% over a year, it was easy to get a buy-in from the hospital, he says. "It was absolutely astounding" data.

The same resolve should start to apply to doctors, as well, Schink adds. As patients start to learn more about the advantages of robotics and demand more minimally invasive procedures, doctors will have to become proficient. "That strikes me as the future of medicine," he says.

St. Joseph's already understands the business case for robotic surgery and views the college as a strategic move forward in surgical services, according to Kirk Wilson, president and CEO of St. Joseph's. In addition, the health system is planning to develop a total of five surgery rooms designed especially for robotic surgery. St. Joseph's is finishing the plans now and needs regulatory approval before starting to build, which is expected to take about a year.

Initially, robotic surgery is not more profitable than traditional surgeries, but the cost per case is starting to decline, Wilson says. "In our view it can only get less costly."

Now You See Me . . .*

By Claire O'Connell
The Irish Times, February 11, 2008

An army of tiny, robotic ants stands motionless before a flickering beam of light, waiting patiently for instructions. Within minutes the 120-strong swarm sets off to fetch and retrieve objects as told, negotiating obstacles in their path as the robo-ants constantly keep track of themselves and each other until their work is done.

Welcome to the world of miniature robotics, where the next generation of tiny, smart devices is being dreamt up and groomed to take on the dull, dirty and downright dangerous jobs of the future.

These miniature workers will be honed to perform a host of tasks. Self-powered, mobile and tuned into their environment, they can wriggle into inaccessible places and report back to base.

And whether it's scurrying off to clean the insides of pipes, hovering around the globe to detect environmental change or swimming inside your body to zap disease, micro- and nano-bots are firing imaginations and bringing powerful technologies together.

But at what cost? While experts dispute the science-fiction accounts of self-assembling nanobots ganging up on us and taking over the earth, there are genuine privacy concerns over where tiny monitoring devices could end up, and that the data they generate could be abused.

That doesn't bother the robo-ants though, who have little space for higher intelligence on their four-millimetre-cubed bodies. Equipped with location sensors, an energy source and an onboard computer, each ant is independent and communicates via infrared with its fellow ants in the swarm, explains Dr Ramon Estaña, a researcher with the Institute for Process Control and Robotics at the University of Karlsruhe in Germany.

The robotic ants are the product of I-swarm, a five-year, pan-European project that has gone back to first principles to recreate swarm behaviour in groups of

* Originally published in *The Irish Times*. Article reprinted with permission of the author.

robots. Biologists studied real ants and informed the software programmers, who faced the challenge of developing complex algorithms for the tiny individuals to act like their living cousins, explains Estaña.

And now that the proof-of-principle model is up and swarming, Estaña believes the tiny robotic insects will find niches in industry over the coming decade. "What we are doing now is basic research but we have visions of cleaning surfaces, going into regions where a normal robot can't move," he says.

"At the moment we do not have the space on the robots but in principle it's possible that they can have sensors to smell, and we have one or two robots which can pick up very small parts with a vibrating needle and transport it to the host."

However, these robo-ants are clunky giants compared with microbots being developed at Monash University in Australia to swim through the human body. They want to make a self-propelling device the width of two human hairs that can shimmy along bloodvessels and help doctors perform minimally invasive treatments. Once more, nature provides the cue and the scientists are developing a motorised tail to allow the microbot to swim, much like the one used by the bacterium E. coli.

Smaller again are the nanobots, on the scale of billionths of metres. It's at this scale that we are most likely to see a rapid impact, particularly in areas such as diagnosing disease and delivering highly selective drugs, according to Dr Diarmuid O'Brien, executive director of Trinity's Centre for Research on Adaptive Nanostructures and Nanodevices (Crann).

But the artificial nanobots that course through our blood vessels in the future are not likely to be miniature versions of conventional robots.

At this level it's more about souped-up molecules built from the bottom up by putting chemicals together to create a new structure. Designer molecules are not how many of us would conventionally think of robots, admits O'Brien, but these complex chemicals can be developed to perform highly specific tasks such as delivering therapies in a smart way by locating a target, communicating its position and carrying out an action.

"Typically you would have a gold nanoparticle that you would functionalise with a protein receptor to recognise a particular cancer cell. So you would release it into the bloodstream and it would find the diseased cell. Then you would use a trigger outside the body to get the gold particle to do something, to kill or mutilate or change the cancer cell."

By assembling bots from specific molecules, scientists are again copying recipes from nature, says O'Brien. "It's taking the best part of traditional medicine and pharmacology and linking it up with some of the possibilities that nanomedicine has provided and creating a smarter and more effective, more efficient, safer drug or diagnostic treatment."

But we are still some time off from seeing nanobot-like drug delivery in use, he adds. "The amount of in vivo testing of this capability has been very limited internationally. Even at this point people are talking about possibilities rather than what's happening on a day-to-day basis in labs or in patient treatment."

The hope is that the nano approach will be a shot in the arm for drug discovery, which has become sluggish over recent years. "In general you see a real slowdown in drug discovery and they are not coming out with the volume and regularity of new treatments as before. There is a sense that this approach with nanotechnology might enable a fresh supply of new ideas, drugs and treatments," says O'Brien.

Tiny robots are also the perfect partners for sensors, and this marriage is set to bring monitoring of both health and the environment to a new level.

Hand-held devices could in future allow us to screen for early stages of disease at home. By washing a drop of blood or saliva over a suite of tiny chemical sensors loaded onto a chip, the device could spot "biomarkers" or signs of trouble and send the results wirelessly to a health-care monitoring service.

Problems can be picked up early and in theory treated before they reach crisis point. Monitors in homes could also detect movement and heartbeat and send out an alert if needed. "We are moving into an aging population base and as people get older and more interdependent, you are creating a smart home to act as a third eye on people," says O'Brien.

Big Brother could swing into action on a wider scale too. Micro- and nanobots loaded with sensors could be deployed to monitor environments around the globe and glean valuable data on climate change on land and in the oceans.

But having sensitive medical data whizzing into databases and the prospect of tiny monitors blinking in the atmosphere are prompting genuine concerns over privacy and data protection, and experts would like to see more public discussion about the social changes that lie ahead as technology develops.

In particular, the military is interested in further miniaturising surveillance devices, notes Prof Dermot Diamond, who directs the National Centre for Sensor Research at Dublin City University.

"At the moment microbots and nanobots are in the realm of hype, but a lot of people are very concerned about how you police this and who owns the information and how is the information stored and used," says Diamond. "You need to have checks and balances in there and make sure that this information, once it's acquired, is used properly and is policed properly. I would take the ethical concerns seriously."

This may seem a giant leap from a bunch of metal ants going about their business, but experts warn of the need to ensure the nascent technology is used ethically.

"It's people's fear of the unknown because we are going into unknown territory and I'm afraid that the track record of human beings using technology in a positive way is not all that good," says Diamond.

IS THIS A ROBOT RISING?

Nanobots are looming on the horizon, but are they going to take over the world? Not likely, according to experts.

We may have fanciful notions from science-fiction of tiny rocket-like robots invading our bodies, or even self-replicating "grey goo" destroying all in its path, but neither is realistic.

The *Fantastic Voyage* like model of a shrunken vehicle just isn't supported by engineering, according to Dr Diarmuid O'Brien of Trinity's Centre for Research on Adaptive Nanostructures and Nanodevices (Crann).

"At the nanoscale those kinds of micro-mechanical machines really just aren't feasible."

Even the bottom-up approach of self-assembling designer molecules has its limitations, and O'Brien dismisses speculation about self-organising nanobots that can take over.

"Molecules behave in a certain way in environmental settings, and applying that in a broader space isn't really credible," he explains.

Much more likely is that social change will creep in on the back of the developing technologies that extend our reach as a species.

As micro- and nanobots improve our capacity to monitor and exploit the earth, store data and detect and treat disease, issues around privacy and healthcare delivery will come to the fore. Because of this, O'Brien calls for more public debate on the matter. Sticking our heads in the sand is of little use, he says.

"From a technological perspective in the 15-20 year timeframe there's really a possibility to use technology to help you manage and order your life in a very different way," he says.

He reckons society will co-adapt as technology develops over the next two decades.

"For example, 20 years ago no one had a mobile phone, and now we now take it for granted, even though many would argue the work-life balance isn't as good because you are more contactable," he says.

"But society has to choose what to accept."

Swallow the Surgeon*

The Economist, September 6, 2008

In the 1966 film "Fantastic Voyage," a submarine carrying a team of scientists is shrunk to the size of a microbe and injected into a dying man. The crew's mission is to save the man's life by dissolving a blood clot deep inside his brain. After a harrowing journey through the patient's body, the scientists succeed: the clot is destroyed and the patient cured.

For decades, scientists and fiction writers alike have been fascinated by the possibility of tiny machines that can enter a patient, travel to otherwise inaccessible regions, and then diagnose or repair problems with far less pain and with far greater precision than existing medical procedures. In his famous speech from 1959, "There's Plenty of Room at the Bottom," Richard Feynman, an American physicist, called this concept "swallow the surgeon." More recently proponents of nanotechnology have imagined swarms of "nanobots"—tiny machines just billionths of a metre, or nanometres, across—that might fix mutations in a person's DNA or kill off cancer cells before they have a chance to develop into a tumour.

Such nanobots still exist only in the realm of science fiction, of course, and it may take decades before they become practical. But there is progress in developing small medical robots for sensing, drug delivery or surgery inside the human body. The most clinically advanced devices focus on the human gastrointestinal (GI) tract, since it is easily accessible and can accommodate objects several centimetres in size. In addition, researchers are developing micro-robots, with dimensions measured in millimetres or micrometres (a human hair is around 100 micrometres in diameter), which should be able to reach more delicate areas, such as the inside of the eyes or the bloodstream.

Among the pioneers of gastrointestinal devices is Given Imaging, an Israeli company that developed the "Pillcam"—a camera-equipped capsule that measures 1.1cm in diameter and 2.6cm in length, and can be swallowed. On its journey through the GI tract the device takes pictures which are transmitted to a com-

* Article from *The Economist*, September 6, 2008.

puter. Because it causes less discomfort and is more effective than other diagnostic methods for examining the small bowel, it has become widely used since its introduction in 2001.

Although the Pillcam is technically not a robot—it is a passive device that relies on the body's regular peristaltic contractions to propel it through the intestine—its success opened up the field, says Paolo Dario, a professor of biomedical robotics at Scuola Superiore Sant'Anna in Pisa, Italy. Research groups around the world are now taking the next steps. At Dr Dario's laboratory, the Centre for Research in Microengineering, scientists have been designing prototypes of capsule robots that have legs for active locomotion. The legs, which are slightly curved, resemble those of insects, with tiny hooks on the tips of their feet to provide enough friction to enable movement in a slippery environment. This should allow the robots to crawl around the GI tract and obtain detailed images, dispense therapeutics or, with the right surgical tools, perform biopsies.

Dr Dario's group is also collaborating with other European researchers on an ambitious project called ARES (short for Assembling Reconfigurable Endoluminal Surgical system). Its objective is to design a modular gastrointestinal robot made up of individual pieces that are small enough to be swallowed, one at a time. Once inside the stomach, the idea is that these pieces will assemble themselves into a larger robotic device. The aim is to build an "operating room" inside a patient that can be controlled from the outside by a doctor, says Dr Dario, who is co-ordinating the project.

He and others in the field acknowledge that there are many obstacles to overcome. A big problem, for example, is how to provide power to tiny medical robots. Batteries can provide enough energy for passive capsules like the Pillcam, but robots with active locomotion pose more of a challenge, and micro-robots are likely to need more energy than batteries can store at such small scales. Instead of adding a power source to the device, which increases its weight and bulk, one approach is to apply external magnetic fields to a small robotic device that contains magnetic material, allowing it to be steered simply by controlling the magnetic fields around it.

Scientists from the Institute of Robotics and Intelligent Systems at the Swiss Federal Institute of Technology (ETH) in Zurich plan to use this technique to steer tiny robots inside the eye for sensing, drug delivery and surgery. Current retinal procedures to repair detachments or rips, for example, may involve several incisions in the eye and stitches to tie off the perforated areas.

Using a micro-robot, by contrast, might involve only one incision and smaller surgical instruments. It might thus require only a mild topical anaesthetic and no stitches, since eye incisions at very small scales can self-seal, says Jake Abbott, a researcher at the institute who is working on the project. In addition, micro-robots could deliver drugs directly to veins inside the retina that have become obstructed by clots, a condition that can cause blindness and for which there is at present no reliable cure. Researchers at the ETH have designed a special ophthalmoscope with which they plan to track the robots inside the eye.

Sylvain Martel and his colleagues at the NanoRobotics Laboratory at Ecole Polytechnique de Montréal in Canada are also using magnetic fields, but in a different way. They are using fields generated by a magnetic-resonance imaging (MRI) machine to ferry small beads through the bloodstream with the goal of delivering therapeutics close to tumours. This has several advantages, says Dr Martel. For one thing, most hospitals already have an MRI machine, so there is no need to construct or buy additional equipment. Furthermore, as well as propelling a magnetic device through the body, an MRI machine can also locate it.

All of this is possible because an MRI machine contains both a large magnet that creates a strong magnetic field, and also a set of gradient coils that superimpose weaker but adjustable fields upon it. These coils are normally used to select slices for creating three-dimensional images, but the magnetic fields they create can also be used to exert a force on a small magnetic object inside a scanner. Dr Martel and his colleagues have developed software to do just that. Last year his team achieved a milestone when it manoeuvred a 1.5-millimetre bead through a 5-millimetre artery in a living pig. Since then the researchers have reduced the bead's size to about 250 micrometres.

But magnetic propulsion has some drawbacks. The fields must be carefully controlled, or they could cause a micro-robot to go off course, with potentially harmful or even fatal consequences inside a human body. And the fields produced by gradient coils in a conventional MRI machine are not strong enough to pull on particles below about 250 micrometres in size, says Dr Martel, though an upgraded MRI system with more powerful coils could propel beads as small as 50 micrometres, he adds. Still, that means smaller blood vessels and other delicate areas inside the body remain out of reach.

To circumvent these problems, some researchers are looking to nature for inspiration. Microbes, for example, have developed successful methods to manoeuvre within bodily fluids. Whereas large animals swim by pushing against water with their fins or limbs, this approach is ineffective at very small scales. To microbes, fluids appear thick and still, and viscosity is the main force small organisms must reckon with. As a result, bacteria have developed a unique way of swimming: using tiny rotary motors called flagella, which resemble corkscrews.

James Friend, co-director of the Micro/Nanophysics Research Laboratory at Monash University near Melbourne, Australia, is building a flagella-inspired micromotor he hopes will one day propel a micro-robot through an artery or vein. At the core of the motor are piezoelectric materials—special crystals or ceramics that change shape very slightly in the presence of an electric field. When such a material is placed in a rapidly alternating electric field, it starts to vibrate. That vibration can then be coupled to another structure to turn a rotor, which in turn operates a flagellum-like tail. In recent years Dr Friend has built successively smaller versions of his motor—the current version is 250 micrometres in diameter. Providing an on-board power supply is difficult, however, so he is investigating the use of external magnetic fields to power the device.

An alternative approach is not to imitate nature, but to harness it—by hitching a ride with a bacterium. To get beyond the size limitations of magnetic beads inside an MRI scanner, Dr Martel and his colleagues are also working with "magnetotactic" bacteria, which orient themselves with magnetic fields. Because they are so tiny (only about two micrometres across), they are not strong enough to swim against the blood flow of larger vessels, though they are able to swim through vessels as little as four micrometres in diameter. Dr Martel's idea is to use the larger magnetic beads to transport the bacteria close to a tumour, and then release them and coax them, using applied magnetic fields, to swim to the tumour and deliver a therapeutic payload. Preliminary experiments in rats suggest that the bacteria can be steered toward tumours using specially designed magnetic coils.

Metin Sitti, director of Carnegie Mellon University's NanoRobotics Lab in Pittsburgh, Pennsylvania, is using bacteria as biological motors to propel small spheres through fluids. Instead of relying on an external system for controlling their movements, Dr Sitti and a colleague use chemical signals to tell the bacteria what to do. In recent experiments they proved that they could stop and start the bacteria's flagella simply by exposing them to two different kinds of substances. Successful steering of the bacteria will be the subject of further tests, says Dr Sitti, but could be done via chemotaxis—a process by which small organisms follow gradients in the concentration of a particular chemical in order to find food or escape from toxic substances.

Medical robots have much potential, but many of the proposed devices, especially those that incorporate biological organisms, may not be practical for many years. Dr Sitti, for one, acknowledges that his project may take a decade or longer to commercialise. But other efforts, such as adding active locomotion to gastrointestinal capsules, may be only a few years away. Dr Dario says medicine is "at the beginning of a new era" as open-wound surgery gives way first to minimally invasive techniques, and then to procedures that will be completely concealed and leave no visible scars. Evidently the ideas depicted in "Fantastic Voyage" continue to resonate: a remake of the film is due to be released in 2010.

Robots*

The Next Generation

By Joseph Mantone
Modern Healthcare, March 21, 2005

Just about every job description includes tasks involving repetition, monotony or drudgery. At healthcare organizations, such work often distracts employees from their main responsibilities and can detract from patient care. One solution: Let robots do more of the chores. Machines are even playing a bigger role in clinical decisionmaking.

It's the latest wave of automation and computerization, and the technologies are being installed at more hospitals nationwide in the name of efficiency and patient safety.

Magee-Women's Hospital in Pittsburgh is among the healthcare organizations that have joined the age of robotics. "Some (employees) like to push the medical record carts around," says Linda Antonelli, vice president of facilities and support services at Magee-Women's. However, she realizes it's certainly not the most efficient use of their time.

In an effort to keep more hospital staffers at the point of care by lessening some of the more mundane tasks, 205-bed Magee-Women's has started using a robot to transport supplies to various parts of the hospital. Meanwhile, in hospital laboratories and pharmacy departments, automated devices also are helping to improve quality by speeding up test results and impeccably dispensing drugs. Robots also are being employed to help improve doctor-patient communication. And computers keep getting smarter—and more robotlike—all the time.

AIMING TO REDUCE ERRORS

Error reduction—an imperative at all hospitals, especially following the landmark 1999 Institute of Medicine report on medical mistakes—is at the core of investment in these technologies. And it's certainly a reason for rising interest in artificial intelligence that is part of decision-support systems designed to aid physicians in treating illness and prescribing medications. The decision-support systems work in conjunction with computerized physician order-entry systems, which have so far been installed at only a small fraction of the nation's hospitals. About 10% of acute-care hospitals have CPOE, according to William Bria, chairman of the Association of Medical Directors of Information Systems. But another organization that tracks the systems, KLAS Enterprises, estimates that only about 4% of hospitals, excluding military and Veterans Health Administration facilities, have operating CPOE systems.

Despite the unimpressive numbers, Bria is a believer in the systems' potential to reduce errors. "There are a lot good products being developed that will be a good extension on what's already out there," he says.

However, he also says many products have been rushed to the market without sufficient research and development, noting that the current systems have experienced limited success. Bria's claim is supported by an article published earlier this month in the *Journal of the American Medical Association* that says CPOE can actually increase the number of medication errors (March 14, p. 10). And the systems aren't cheap. Bria says the most complex CPOE systems for large academic hospitals can cost as much as $40 million.

One reason Bria isn't enamored with current systems is because they are equipped with what he calls "dumb alerts," which warn physicians when they've nearly completed a prescription order. "We need alerts before decisions are made," he says.

Joel Diamond, a private family-practice physician who works at 185-bed UPMC St. Margaret Hospital in Pittsburgh, helped the hospital select its CPOE system. Diamond doesn't have supporting data on the system—in use at St. Margaret since September 2004—but he says there's no doubt care has improved. One way it has helped, Diamond says, is by reducing the number of tasks physicians forget to do, such as placing orders for tests or pharmaceuticals.

Diamond admits that physicians at St. Margaret have been ignoring some alerts—the doctors contend some of the prompts are often false alarms—but he says improved compliance is just a matter of fine-tuning the system. "This technology is advancing like cell phone technology," he says. "The cell phone you have today won't be the same cell phone you have in the future."

He adds that he wasn't too surprised with the JAMA findings because users always make errors when dealing with a new technology.

DELIVERING RESULTS

Also hard at work in more of the nation's hospitals are courier robots, as well as robotic communication devices that facilitate doctor-patient interaction. Courier robots are newer technologies, and it's difficult to pinpoint how much hospitals are investing in them overall. At least three companies—Aethon; Pyxis Corp., a subsidiary of Cardinal Health; and California Computer Research—offer the machines.

The Aethon product, called a Tug, is being used in about six hospitals including Magee-Women's, which has been a testing site for the robot since 2003. The company charges about $1,250 per month for the Tug, but an Aethon cost analysis shows that hospitals on average would pay a person $92.50 per shift to do what a Tug can do for about $18.86.

Robert Brennan, director of materials management at Magee-Women's, says the Tug has been able to help reduce the hospital's overtime costs but it's too early to come up with a firm figure. The hospital uses two Tugs, one with a locked cabinet for transporting pharmaceuticals and another with an attached cart that transports medical equipment, usually from the supply room to the surgical suite.

The machine is about a foot long and about half a foot high and attaches to the bottom of a specially designed cart. It has a built-in sensor, so if someone or something does get in its way, it will stop before a collision occurs. The Tug is also equipped with a camera, which beams images to Aethon's headquarters, where operators monitor the Tugs round the clock. Workers can direct the robots from Aethon's office or the hospital.

A map of the hospital is programmed into the Tug. It usually works like this: When a delivery request is made, a supply-room worker will load the cart and then select the Tug's destination through a touch screen. On the screen, workers can track its progress. As the machine moves about the hospital, it triggers sensors that open doors and summon elevators. As it approaches an elevator, the Tug will announce, "Calling elevator."

Speakers are located inside the rooms where the Tug makes deliveries. Once the robot reaches its destination it sends a message—which is audible through the speakers—alerting workers to unload the cart. After unloading, the Tug then returns to the supply room, where it can be restocked or insert itself into a charger.

Now, many in the hospital would like to have a Tug in their departments—indeed, a hospital lawyer jokes that she wants a paralegal Tug—but Antonelli says departments can add them only if their purchase is budget-neutral.

Cardinal Health purchased HelpMate Robotics, which developed a courier robot in 1999. The HelpMate, which stands 4 feet 6 inches tall, can weigh up to 600 pounds and has a 24-inch-by-26-inch storage capacity. The company has deployed about 108 robots at 78 healthcare facilities, and like the Tug they're used to make deliveries of various supplies and equipment. It differs from the Tug because the entire robot is one unit, while the Tug attaches itself to a cart. Including installation, the HelpMate goes for about $150,000—it can also be rented—and the company says a hospital can see a return on investment in 18 to 24 months.

Unlike the Tug or HelpMate, California Computer Research's RoboCart runs along a fixed path. The RoboCart is mostly used to transport specimens from one end of a clinical lab to the other. The model is being used in about 10 hospitals in the U.S. and in 15 abroad, according to its manufacturer. At a RoboStation, robotic arms attached to a table load and unload the RoboCart. The RoboCart travels along a path marked by special tape, and technicians can operate it with a hand-held device, according to its manufacturer. The RoboCart costs about $1,375 per month to rent or about $25,000 for purchase and installation, according to the company.

Denise Geiger, laboratory director at 248-bed John T. Mather Memorial Hospital in Port Jefferson, N.Y., thought about adding a courier robot to address some of the workforce shortage problems at her hospital, but instead she decided to purchase a lab automation system. The hospital has witnessed a 100% increase in emergency room visits since 1996—attributed to a growing and aging population in its market—and now receives about 40,000 visits per year.

DOUBLE THE VOLUME

Despite the accompanying increase in lab workload, the hospital hasn't had to add staff, in part thanks to increased automation. "We have the same number of (full-time equivalent staff) and double the volume," Geiger says.

About four years ago, the hospital bought a lab automation system from Beckman Coulter. These systems, which move blood samples like bottles through a brewery, are currently in about 200 hospitals, but that number is expected to reach about 300 hospitals by year-end, according to *Diagnostic Testing & Technology Report*, an industry newsletter.

At Geiger's hospital, about 60 tests can be performed through the system, and it takes about 45 to 60 minutes for completion. "Before it was unpredictable" when a test might be completed, she says. The results have improved patient flow, which leads to fewer unnecessary admissions and savings of close to $2 million per year, according to the hospital.

The process usually starts in the emergency room. ER staff members label a lab specimen with a bar code that indicates which patient the sample is from and what tests need to be performed. The samples are then put on a conveyor belt that leads into the lab. From there, the sample can be tested and placed into a refrigerated storage bin, all without a lab technician handling it. This process eliminated about 14 steps for technicians, and it's helped the hospital realize a 20% decrease in possible errors, Geiger says.

McKesson Corp.'s Robot-Rx automates a hospital's pharmacy department by using a roughly 7-foot robot that moves along a fixed path in the centralized drug distribution system. Bar-coded medications are stored on shelves and orders are sent to Robot-Rx electronically.

The system pulls the medications off the shelves, labels the proper patients' information and dosage with a bar code and drops the medication into a bin. From there, nurses use a hand-held device to scan the bar code and ensure it's the proper medication before administering it.

Children's Hospital Central California, a 246-bed facility in Madera, has been using Robot-Rx for three years and the $1 million system has helped lower medication error rates to less than 1%, says spokeswoman Micheline Golden. Last fall David Brailer, HHS' health information technology czar, saw the Robot-Rx at work at 1,285-bed UPMC Presbyterian in Pittsburgh, which has been using the system since 2001.

THE ROBOT IS IN

To support another basic task at hospitals—communication between patients and caregivers—InTouch Health has developed a 5-foot-6-inch mobile robot that might stir up images of Number 5 from the movie "Short Circuit." Unlike Number 5, this robot must be operated by a user.

The robot is designed for doctors to communicate with patients, even when doctors are not in the same building, a process that's been dubbed "telerounding." The University of California-Davis Medical Center took part in a six-month study that examined whether telerounding is as safe as performing traditional rounds. "It's basically a remote teleconferencing system," says Charles Casey, a hospital spokesman. Hospitals need a wireless Internet connection to use the robot.

Lars Ellison, a UC-Davis urologist, used the robot, which the hospital nicknamed Rudy, in the study that recently concluded but doesn't yet have the results. When using Rudy, Ellison could sit in his office, located about three blocks from the hospital, and check on patients. A camera pointed at the doctor sends his image to the screen that serves as Rudy's head. Ellison speaks into a microphone and steers Rudy with a joystick.

The hospital leased the robot for about $3,000 per month, and UC-Davis is considering using Rudy as a teaching tool and maybe even a translator. The hospital hasn't identified how Rudy can save money, but Casey says the benefits will be in improved patient satisfaction.

The robot is good for doctors such as Ellison, who work at more than one hospital. It's often more reassuring for patients to speak with a doctor who has been treating them, and the camera on Rudy lets him zoom in close, enabling the doctor to examine an incision.

"There's a chuckle factor involved," Casey says. He adds: "On its surface it's strange, but everyone has liked it."

Paging Dr. C-3PO*

By Dave Carpenter
Hospitals & Health Networks, July 2002

What's 12 feet tall, fast and makes no errors? No, not Shaquille O'Neal's brother. It's Sammy, Alegent Health's newest helper in the pharmacy, who dispenses inpatient medications and could be a slam dunk for patient safety.

Sammy—really SAMI (Safe Accurate Medications for Inpatients), Alegent's new pharmacy robot—came on duty in late April at Bergan Mercy Medical Center, Omaha, Nebr. It began by dispensing nearly 4,000 doses of medications daily, a fraction of its capacity of 15,840 doses when fully loaded. Eventually, SAMI will handle all inpatient medications at Alegent's four Omaha-area hospitals, setting it apart from most other pharmacy robots, health system officials say. Only SAMI, officials contend, can serve an entire health system.

Dispensing robots are slowly becoming more popular at U.S. hospitals. Also getting off the ground are mobile robots, R2D2-like machines that deliver medications from the pharmacy to the nurse station nearest the patient's room; nearly 100 such machines by Pyxis Corp., San Diego, are in use nationwide, including at seven Veterans Administration hospitals.

At Bergan Mercy, staff liken SAMI in appearance to a giant jukebox, spinning medications instead of music. It operates with 11 mechanical rings, which each contain 72 spirals; on each spiral are 20 grooves that each hold a medication.

That all adds up to speed and efficiency, says Donald Manning, M.D., Alegent's chief medical officer. "What used to take pharmacists about six hours—an overnight fill of the next day's prescription—with the chance that there could be errors made, now takes an hour and a half and it's flawless," Manning says.

The refined safety process begins when the robot packages bulk medications into individual packets bearing bar codes containing essential information. Packets are stored in the spiral-and-groove system, requested and confirmed by scanning the bar code, and dispensed onto conveyor belts that empty them into a drawer. "The bar code scanning prevents the human error that can occur when the medications are manually picked out of bins," says Angela Ward, operations director for pharmacy. "Studies have shown that 11 percent of medication errors occur when the pharmacy dispenses a medication. That won't happen with this system."

SAMI, a Homerus Robot by Pyxis, is more expensive than the average pharmacy worker, retailing for about $1 million—the vendor won't provide the exact price—and costing $4,000 a month to operate. (McKesson Corp., San Francisco, makes a competing version called ROBOT-Rx; another competitor is ScriptPro, Mission, Kans.) Yet Alegent sees SAMI providing savings, including bulk medication prices that could help cancel out the monthly operating costs. A more tangible benefit in a time of workforce shortages: pharmacists are freed up to focus on patient care, interacting more with patients and consulting more with clinicians. "It is an absolutely fantastic concept," Manning says. Not only that, it looks cool, too.

4

Robots in Space

Editor's Introduction

In the wake of the February 2003 Space Shuttle *Columbia* disaster—a tragedy that claimed the lives of seven astronauts in the skies over Texas—the National Aeronautics and Space Administration (NASA) has faced increased scrutiny from those who question the safety, not to mention the value, of sending humans into space. Indeed, manned space missions are dangerous and expensive, and it's partially for these reasons that some have urged NASA to shift its focus to robots—autonomous rovers, orbiters, and probes that have already shown great promise.

Since 1997, NASA has been using robots to explore the surface of Mars, a place where man has yet to set foot. While the first Martian rover, Sojourner, only managed to travel 100 meters, subsequent robots Spirit and Opportunity have, as of September 2009, combined to cover more than 24 kilometers. Increasingly, technology has enabled probes and rovers to function autonomously and gather data without having to rely on instructions from ground control. In 1999, the Deep Space I probe journeyed more than 600 kilometers and steered itself onto an asteroid. Closer to home, the Earth Observing-1 satellite, or EO-1, uses an autonomous control system to scan the planet for floods and volcanic eruptions.

The articles in this chapter provide an overview of how robots are being used in space exploration. The selections also highlight efforts to make robots function more effectively and independently in the future. In the first article, "Space Robot 2.0," Molly Bentley explains how software helps Martian rovers "know" how to ration battery power and concentrate on photographing only those objects that are likely to be of interest to scientists.

In "Robonauts," the second selection, Carolyn Y. Johnson considers how robots might partner with—not replace—humans on future space missions. The article's title refers to a robot that is being designed to manipulate tools and complete space walks. "The danger is sticking with the mind-set that developed in the 1960s of 'what robots do' and 'what humans do,'" Rodney Brooks, the director of the Computer Science and Artificial intelligence Laboratory at the Massachusetts Institute of Technology (MIT), tells Johnson.

The next piece, "NASA and Caltech Test Steep-Terrain Rover," centers on "Axel," a prototype rover whose low mass allows for the exploration of cliffs and rocky terrain. The result of a partnership between NASA and the California Institute of Technology (CalTech), Axel boasts a levered arm for collecting soil samples and a pair of cameras that rotate 360 degrees. The subsequent selection, "Robot Armada Might Scale New Worlds," looks at how teams of robots—ground rovers, low-flying balloons, and high-altitude orbiters—might gather and transmit back to Earth information on moons and planets. "This armada of robots will be our eyes, ears, arms and legs in space, in the air, and on the ground, capable of responding to their environment without us, to explore and embrace the unknown," Wolfgang Fink, director of Caltech's Visual and Autonomous Exploration Systems Research Laboratory, tells the author.

The next entry in this chapter, "NASA to Begin Attempts to Free Sand-Trapped Mars Rover," details efforts by scientists to maneuver Spirit out of a hole it's been stuck in since April 2009. It remains unclear whether the rover—which, even before getting trapped, was rolling around with a broken wheel—will make it out of its predicament, though scientists are viewing the operation as win-win. "Mobility on Mars is challenging, and whatever the outcome, lessons from the work to free Spirit will enhance our knowledge about how to analyze Martian terrain and drive future Mars rovers," says Doug McCuistion, director of the NASA's Mars Exploration Program.

In the final piece, "Robot Firm Aims for Contract to Build Lunar Launch Pad," David Templeton writes of how the firm Astrobotic Technology, Inc. hopes to use robots to build a launch pad and landing facility on the moon. One of the major challenges associated with such a project, Templeton explains, is protecting people and equipment from moon dust, a "fine, spiky, self-cohesive" substance that tends to work its way into instruments and foul up lunar experiments.

Space Robot 2.0*

Smarter than the Average Rover

By Molly Bentley
New Scientist, September 14, 2009

Something is moving. Two robots sitting motionless in the dust have spotted it. One, a six-wheeled rover, radios the other perched high on a rocky slope. Should they take a photo and beam it back to mission control? Time is short, they have a list of other tasks to complete, and the juice in their batteries is running low. The robots have seconds to decide. What should they do?

Today, mission control is a mere 10 metres away, in a garage here at NASA's Jet Propulsion Laboratory (JPL) in Pasadena, California. Engineers can step in at any time. But if the experiment succeeds and the robots spot the disturbance and decide to beam the pictures back to base, they will have moved one step closer to fulfilling NASA's vision of a future in which teams of smart space probes scour distant worlds, seeking out water or signs of life with little or no help from human controllers.

NASA, along with other space agencies, has already taken the first tentative steps towards this kind of autonomous mission. In 1999, for example, NASA's Deep Space 1 probe used a smart navigation system to find its way to an asteroid—a journey of over 600 million kilometres. Since 2003, an autonomous control system has been orbiting our planet aboard NASA's Earth Observing-1 satellite. It helps EO-1 to spot volcanic eruptions and serious flooding, so the events can be photographed and the images beamed back to researchers on the ground. And in the next month or so, the latest iteration of smart software will be uploaded onto one of NASA's Mars rovers, loosening the machine's human tether still further so it can hunt for unusual rock formations on its own.

The idea is not to do away with human missions altogether. But since it is far cheaper and easier to send robots first, why not make them as productive as possible? Besides, the increasingly long distances they travel from home make controlling a rover with a joystick impractical. Commands from Earth might take 20 minutes to reach Mars, and about an hour to reach the moons of Jupiter.

So what can we realistically expect autonomous craft to do? It is one thing to build a space probe that can navigate by itself, respond quickly to unexpected events or even carry on when a critical component fails. It's quite another to train a planetary rover to spot a fossilised bone in a rock, let alone distinguish a living cell from a speck of dirt.

The closest thing to a space robot with a brain is NASA's pair of Mars rovers, and their abilities are fairly limited. Since they landed in January 2004 they have had to cope with more than six critical technical problems, including a faulty memory module and a jammed wheel. That the craft are still trundling across the red planet and returning valuable geological data is down to engineers at mission control fixing the faults remotely. In fact the rovers can only do simple tasks on their own, says Steve Chien, the head of JPL's artificial intelligence group. They can be programmed to drive from point A to point B, stop, and take a picture. They can spot clouds and whirling mini-tornadoes called dust devils on their own. They can also protect themselves against accidental damage—by keeping away from steep slopes or large rocks. For pretty much everything else, they depend on their human caretakers.

WHAT ARE WE MISSING?

This is becoming a significant limitation. While NASA's first Mars rover, Sojourner, travelled just 100 metres during its mission in 1997, Spirit and Opportunity have covered over 24 kilometres so far. As they drive they are programmed to snap images of the landscape around them, but that doesn't make for very thorough exploration. "We are travelling further and further with each rover mission," says Tara Estlin, senior computer scientist and one of the team developing autonomous science at JPL. "Who knows what interesting things we are missing?"

NASA wouldn't want the rovers to record everything they see and transmit it all back to Earth; the craft simply don't have the power, bandwidth and time. Instead, the team at JPL has spent around a decade developing software that allows the rovers to analyse images as they are recorded and decide for themselves which geological features are worth following up. Key to this is a software package called OASIS—short for on-board autonomous science investigation system.

The idea is that before the rovers set out each day, controllers can give OASIS a list of things to watch out for. This might simply be the largest or palest rock in the rover's field of view, or it could be an angular rock that might be volcanic. Then whenever a rover takes an image, OASIS uses special algorithms to identify any rocks in the scene and single out those on its shopping list (*Space Operations Communicator*, vol 5, p39). Not only is OASIS able to tell the rovers what features are of scientific interest, it knows their relative value too: smooth rocks which may have been eroded by water might take priority over rough ones, say. This helps the rovers decide what to do next.

There are also practical considerations to take into account. As they trundle around the surface, the rovers must keep track of whether they have enough time, battery power and spare memory capacity to proceed. So the JPL team has also created a taskmaster—software that can plan and schedule activities. With science goals tugging at one sleeve and practical limitations at the other, this program steps in to decide how to order activities so that the rover can reach its goals safely, making any necessary scheduling changes along the way. With low-priority rocks close by, say, a rover might decide it is worth snapping six images of them rather than one of a more interesting rock a few metres away, since the latter would use up precious battery juice.

Why stop there? Since OASIS allows a rover to identify high-priority targets on its own, the JPL team has decided to take the next step: let the rover drive over to an interesting rock and deploy its sensors to take a closer look. To do this, Estlin and her colleagues won't be using OASIS, however. Instead, they have taken elements from it and used them to create a new control system called Autonomous Exploration for Gathering Increased Science (AEGIS). This has been tested successfully at JPL and is scheduled for uplink and remote installation on the rover Opportunity sometime in September.

Once AEGIS is in control, Opportunity will be able to deploy its high-resolution camera automatically and beam data back to Earth for analysis—the first time autonomous software has been able to control a craft on the surface of another world. This is just the beginning, says Estlin. For example, researchers at JPL and the Wesleyan University in Middletown, Connecticut, have developed a smart detector system that will allow a rover to carry out a basic scientific experiment on its own. In this case, its task will be to identify specific minerals in an alien rock.

The detector consists of two automated spectrometers controlled by "support vector machines"—relatives of artificial neural networks—of a kind already in use aboard EO-1. The new SVM system uses the spectrometers to take measurements and then compares the results with an on-board database containing spectra from thousands of minerals. Last year the researchers published results in the journal *Icarus* (vol 195, p 169) showing that in almost all cases, even in complex rock mixtures, their SVM could automatically spot the presence of jarosite, a sulphate mineral associated with hydrothermal springs.

ALIEN NOVELTIES

Though increasingly sophisticated, these autonomous systems are still a long way from the conscious machines of science fiction that can talk, feel and recognise new life forms. Right now, Chien admits, we can't even really program a robot for "novelty detection"—the equivalent of, say, picking out the characteristic shape of a bone among a pile of rocks—let alone give it the ability to detect living creatures.

In theory, the shape of a complex natural object such as an ice crystal or a living cell could be described in computer code and embedded in a software library. Then the robot would only need a sensor such as a microscope with sufficient magnification to photograph it.

In fact identifying a cell is a huge challenge because its characteristics can be extremely subtle. In 1999, NASA funded an ambitious project that set out to discover whether there are specific signatures such as shape, symmetry, or a set of combined features that could provide a key to identifying and categorising simple living systems (*New Scientist*, 22 April 2000, p 22). The idea was to create a huge image library containing examples from Earth, and then teach a neural network which characteristics to look for. Unfortunately, the project ended before it could generate any useful results.

Just as a single measurement is unlikely to provide definitive proof of alien life, so most planetary scientists agree that a single robotic explorer, however smart, won't provide all the answers. Instead, JPL scientists envisage teams of autonomous craft working together, orbiting an alien world and scouring the surface for interesting science, then radioing each other to help decide what features deserve a closer look.

This model is already being put through its paces. Since 2004, networks of ground-based sensors placed around volcanoes, from Erebus in Antarctica to Kilauea and Mauna Loa in Hawaii, have been watching for sudden changes that might signal an eruption. When they detect strong signals, they can summon EO-1, which uses its autonomous software planner to schedule a fly-past. The satellite then screens the target area for clouds, and if skies are clear, it records images, processes them and transmits them to ground control.

In July, a network of 15 probes were placed into Mount St Helens, a volcano in Washington state. These probes carry sensors that monitor conditions inside the crater and can talk to each other to analyse data in real time, as well as call up EO-1 to take photos. If it detects activity from orbit, the satellite can even ask the probes to focus attention on a particular spot.

Networks of autonomous probes can provide a number of advantages, including helping a mission cover more ground, and ensuring it continues even if one or more probes are damaged or destroyed. This approach also offers increased processing power, since computers on separate probes can work together to crunch data more quickly. And researchers are beginning to believe that teams of autonomous probes could eventually be smart enough to do almost everything a human explorer could, even in the remotest regions of space.

Last year, in a paper published in the journal *Planetary and Space Science* (vol 56, p 448), a consortium of researchers from the US, Italy and Japan laid out their strategy for searching out life using autonomous craft controlled by fuzzy logic, the mathematical tool developed in the 1960s to give computers a way to handle uncertainty. Their plan calls for the use of three types of craft: surface-based rovers with sensors designed to spot signs of water and potential sources of heat, such as geothermal vents; airships that float low overhead and help pinpoint the best sites for study; and orbiters that image the planet surface, coordinating with mission control as well as beaming data back to Earth.

The consortium argue that fuzzy logic is a better bet than neural networks or other artificial intelligence techniques, since it is well suited to handling incomplete data and contradictory or ambiguous rules. They also suggest that by working together, the three types of probes will have pretty much the same investigative and deductive powers as a human planetary scientist.

Experimental simulations of a mission to Mars seem to confirm this view: in two tests the autonomous explorers came to the same conclusions as a human geoscientist. The system could be particularly useful for missions to Titan and Enceladus, the researchers suggest, since autonomy will be a key factor for the success of a mission so far from Earth.

Back at JPL, the day's test of robot autonomy is almost complete. The two robots are running new software designed to improve coordination between craft. Part of the experiment is to see whether the robots can capture a photo of a moving target—in this case a small remote-controlled truck nicknamed Junior—and relay it back to "mission control" using delay-tolerant networking, a new system for data transfer.

In future deep-space missions, robots will need autonomy for longer stretches since commands from Earth will take an hour or so to reach them. And as planets rotate, there will be periods when no communication is possible. Delay-tolerant networking relies on a "store and forward" method that promises to provide a more reliable link between planetary explorers and mission control. Each node in the network—whether a rover or an orbiter—holds on to a transmission until it is safe to relay it to the next node. Information may take longer to reach its destination this way, but it will get there in the end.

And it seems to work: the images from the two robots arrive. They include both wide-angle shots and high-resolution close-ups of Junior. Estlin is pleased.

As we stand in the heat, a salamander scuttles quickly across a rock. I can't help wondering whether the robots would have picked that out. Just suppose the Mars rover had to choose between a whirling dust devil and a fleeing amphibian? Chien assures me that the software would direct the rover to prioritise, depending on the relative value of the two. I hope it goes for the salamander. And if alien life proves half as shy, I hope the rover can act fast.

Robonauts*

The Next Generation of Space Explorers Will Look—and Act—More Like People than Probes

By Carolyn Y. Johnson
The Boston Globe, January 9, 2006

In 1989, using an insect-like robot named Genghis, Rodney Brooks pitched a bold vision for exploring space: Send up an army of small, cheap machines to rove around on a distant planet and beam back data.

The concept kicked off a new era in robotics, and eight years later, NASA sent the simple probe Sojourner rolling across the surface of Mars.

But now Genghis sits in a box, and the sophisticated machines that populate Brooks's lab at MIT are becoming increasingly human-like: One has a sense of touch, another can find a familiar face in a crowd. Eventually a robotic torso named Domo—now learning to wield a screwdriver—will be able to master new skills by imitating people instead of undergoing software updates.

The new designs are part of a broader shift toward a vision of robots that are partners, not simply remote-controlled probes.

The change has been fueled by more powerful computers and better robotics as well as by new space policy. The Bush administration's push for more human space flight—signed off on a few weeks ago by Congress—is increasing the demand for robot partners that can learn new tasks, use tools the same way people do, and act as a space support staff.

"The thing we were tasked by NASA is: How can robots support manned missions on the moon and Mars before people get there, while they are up there, and after they've left?" said Brooks, director of the Computer Science and Artificial Intelligence Laboratory at the Massachusetts Institute of Technology. "The danger is sticking with the mind-set that developed in the 1960s of 'what robots do' and 'what humans do.'"

It is now clear that both humans and robots have their advantages in space—and the segregation between the two is fading as NASA pursues colonization of the moon and Mars.

Robots pose less risk: No one loses a parent, child, or spouse when a spacecraft carrying a robot explodes. They are also cheaper and less delicate explorers than their fleshy creators.

The Mars rovers, Spirit and Opportunity, for example, have cost roughly $900 million over 5½ years, according to Joy Crisp, a project scientist at the Jet Propulsion Laboratory in Pasadena, Calif. In contrast, in its first 112 shuttle flights the three-decade-old space shuttle program has cost 14 lives and roughly $1.3 billion per flight—not a great "bang for your buck," said Roger Pielke Jr., director of the Center for Science and Technology Policy Research at the University of Colorado at Boulder.

But even the best-designed robots can't compete with the human brain.

The rovers are an amazing success: Both are still working after two years instead of the projected 90 days. They've each put about 4 miles on the odometer, sent back possible evidence of water, and collected data to keep scientists busy for years.

They were never really alone, though. A ground team of about 50 scientists and engineers told them what to do each day. People mined everything the rovers saw, felt, and measured, and then had to tell them to go back to probe the interesting spots. A person could walk the same four miles in a single day, said Jeffrey Hoffman, a former astronaut who worked to repair the Hubble telescope in 1993.

So the real trick will be to blend the versatility of the human brain with the efficiency of the robotic mechanisms.

"Let them do what they do best, look at it as a team," said Dava Newman, a professor of aerospace engineering at MIT, who studies the biology of humans in space. "You're right there with your robot assistant—the robot is kind of the laboratory, helping analyze the soil, maybe helping carry" the equipment. Newman, like other scientists who believe that people should explore and live in space, realizes that they will depend on machines as precursors and as complicated tools.

In Brooks's lab, robots are developing the skills they'll need to be useful to people—though he still believes that "fast, cheap, and out-of-control" robots will have a role in the future. The personable, twitchy robot Mertz recognizes faces and distinguishes one person from another. Obrero, a mechanical arm, has a touch so sensitive it can pick up a fluffy stuffed cow or a plastic toy—delicate tasks that have foiled many other robots.

At the Johnson Space Center in Houston, researchers are developing Robonaut, an agile, tool-using robot-astronaut that can outlast any human on a space walk. SCOUT, a lunar rover being developed by NASA, will carry astronauts but will also have the potential to act on its own.

Last month, NASA launched two competitions to encourage the private sector to create autonomous robots—ones that can assemble structures with minimal human intervention and ones that can steer along a flight path and touch down to take surface samples.

The urge to make robots more "human" isn't just about sending them to space. Robots will eventually wheel around hospitals, schools, and homes—and they will need to be able to read social cues, learn a task, and be able to do work without supervision.

The latest version of Honda Motor Co.'s ASIMO robot debuted last month in Japan. The 4-foot-tall robot can push a cart and run like a person. Such a human-like robot would be "like having an assistant, a sous chef in a kitchen who might not be as good as the chef, who might not be able to do as much work as the people," said former astronaut Hoffman. "With robotic assistance, people will be much more useful, and, conversely, by having people around, robots will be much more useful."

NASA and Caltech Test Steep-Terrain Rover*

NASA Jet Propulsion Lab, February 4, 2009

Engineers from NASA's Jet Propulsion Laboratory and students at the California Institute of Technology have designed and tested a versatile, low-mass robot that can rappel off cliffs, travel nimbly over steep and rocky terrain, and explore deep craters.

This prototype rover, called Axel, might help future robotic spacecraft better explore and investigate foreign worlds such as Mars. On Earth, Axel might assist in search-and-rescue operations.

A Web video showing an Axel test-run at the JPL Mars yard is online at: www.jpl.nasa.gov/video/index.cfm?id=806 .

"Axel extends our ability to explore terrains that we haven't been able to explore in the past, such as deep craters with vertically-sloped promontories," said Axel's principal investigator, Issa A.D. Nesnas, of JPL's robotics and mobility section. "Also, because Axel is relatively low-mass, a mission may carry a number of Axel rovers. That would give us the opportunity to be more aggressive with the terrain we would explore, while keeping the overall risk manageable."

The simple and elegant design of Axel, which can operate both upside down and right side up, uses only three motors: one to control each of its two wheels and a third to control a lever. The lever contains a scoop to gather lunar or planetary material for scientists to study, and it also adjusts the robot's two stereo cameras, which can tilt 360 degrees.

Axel's cylindrical body has computing and wireless communications capabilities and an inertial sensor to operate autonomously. It also sports a tether that Axel can unreel to descend from a larger lander, rover or anchor point. The rover can use different wheel types, from large foldable wheels to inflatable ones, which help the rover tolerate a hard landing and handle rocky terrain.

Nesnas co-leads the project with Joel Burdick, a mechanical and bioengineering professor at Caltech, who supervises a handful of Caltech graduate and undergraduate students working on the rover system. Last fall, the JPL-Caltech team

---

* Courtesy of NASA.

demonstrated Axel at the annual Smithsonian Folklife Festival in Washington, which showcased NASA for the agency's 50th anniversary.

"Collaboration with Caltech has been key to the success of this project," Nesnas said. "The students contributed significantly to the design of the tethered Axel. Their creative work enabled us to analyze, design and build new wheels, sampling tools and software. The students also played a key role in field-testing this robot. Without them, we would not have been able to accomplish such goals, given our limited resources."

JPL began developing Axel in 1999, in partnership with Purdue University, West Lafayette, Ind., and Arkansas Tech University, Russellville, Ark. The Axel project was funded through NASA's Exploration System Mission Directorate. Caltech manages JPL for NASA.

Robot Armada Might Scale New Worlds*

NASA Jet Propulsion Lab, October 27, 2009

An armada of robots may one day fly above the mountain tops of Saturn's moon Titan, cross its vast dunes and sail in its liquid lakes.

Wolfgang Fink, visiting associate in physics at the California Institute of Technology in Pasadena, says we are on the brink of a great paradigm shift in planetary exploration, and the next round of robotic explorers will be nothing like what we see today.

"The way we explore tomorrow will be unlike any cup of tea we've ever tasted," said Fink, who was recently appointed as the Edward and Maria Keonjian Distinguished Professor in Microelectronics at the University of Arizona, Tucson. "We are departing from traditional approaches of a single robotic spacecraft with no redundancy that is Earth-commanded to one that allows for having multiple, expendable low-cost robots that can command themselves or other robots at various locations at the same time."

Fink and his team members at Caltech, the U.S. Geological Survey and the University of Arizona are developing autonomous software and have built a robotic test bed that can mimic a field geologist or astronaut, capable of working independently and as part of a larger team. This software will allow a robot to think on its own, identify problems and possible hazards, determine areas of interest and prioritize targets for a close-up look.

The way things work now, engineers command a rover or spacecraft to carry out certain tasks and then wait for them to be executed. They have little or no flexibility in changing their game plan as events unfold; for example, to image a landslide or cryovolcanic eruption as it happens, or investigate a methane outgassing event.

"In the future, multiple robots will be in the driver's seat," Fink said. These robots would share information in almost real time. This type of exploration may one day be used on a mission to Titan, Mars and other planetary bodies. Current proposals for Titan would use an orbiter, an air balloon and rovers or lake landers.

In this mission scenario, an orbiter would circle Titan with a global view of the moon, with an air balloon or airship floating overhead to provide a birds-eye view of mountain ranges, lakes and canyons. On the ground, a rover or lake lander would explore the moon's nooks and crannies. The orbiter would "speak" directly to the air balloon and command it to fly over a certain region for a closer look. This aerial balloon would be in contact with several small rovers on the ground and command them to move to areas identified from overhead.

"This type of exploration is referred to as tier-scalable reconnaissance," said Fink. "It's sort of like commanding a small army of robots operating in space, in the air and on the ground simultaneously."

A rover might report that it's seeing smooth rocks in the local vicinity, while the airship or orbiter could confirm that indeed the rover is in a dry riverbed—unlike current missions, which focus only on a global view from far above but can't provide information on a local scale to tell the rover that indeed it is sitting in the middle of dry riverbed.

A current example of this type of exploration can best be seen at Mars with the communications relay between the rovers and orbiting spacecraft like the Mars Reconnaissance Orbiter. However, that information is just relayed and not shared amongst the spacecraft or used to directly control them.

"We are basically heading toward making robots that command other robots," said Fink, who is director of Caltech's Visual and Autonomous Exploration Systems Research Laboratory, where this work has taken place.

"One day an entire fleet of robots will be autonomously commanded at once. This armada of robots will be our eyes, ears, arms and legs in space, in the air, and on the ground, capable of responding to their environment without us, to explore and embrace the unknown," he added.

NASA to Begin Attempts to Free Sand-Trapped Mars Rover*

NASA Jet Propulsion Lab, November 12, 2009

NASA will begin transmitting commands to its Mars exploration rover Spirit on Monday as part of an escape plan to free the venerable robot from its Martian sand trap.

Spirit has been lodged at a site scientists call "Troy" since April 23. Researchers expect the extraction process to be long and the outcome uncertain based on tests here on Earth this spring that simulated conditions at the Martian site.

"This is going to be a lengthy process, and there's a high probability attempts to free Spirit will not be successful," said Doug McCuistion, director of the Mars Exploration Program at NASA Headquarters in Washington. "After the first few weeks of attempts, we're not likely to know whether Spirit will be able to free itself."

Spirit has six wheels for roving the Red Planet. The first commands will tell the rover to rotate its five working wheels forward approximately six turns. Engineers anticipate severe wheel slippage, with barely perceptible forward progress in this initial attempt. Since 2006, Spirit's right-front wheel has been inoperable, possibly because of wear and tear on a motor as a result of the rover's longevity.

Spirit will return data the next day from its first drive attempt. The results will be assessed before engineers develop and send commands for a second attempt. Using results from previous commands, engineers plan to continue escape efforts until early 2010.

"Mobility on Mars is challenging, and whatever the outcome, lessons from the work to free Spirit will enhance our knowledge about how to analyze Martian terrain and drive future Mars rovers," McCuistion said. "Spirit has provided outstanding scientific discoveries and shown us astounding vistas during its long life on Mars, which is more than 22 times longer than its designed life."

In the spring, Spirit was driving backward and dragging the inoperable right front wheel. While driving in April, the rover's other wheels broke through a crust on the surface that was covering a bright-toned, slippery sand underneath. After a

* Courtesy of NASA.

few drive attempts to get Spirit out in the subsequent days, it began sinking deeper in the sand trap. Driving was suspended to allow time for tests and reviews of possible escape strategies.

"The investigations of the rover embedding and our preparations to resume driving have been extensive and thorough," said John Callas, project manager for Spirit and Opportunity at NASA's Jet Propulsion Laboratory in Pasadena, Calif. "We've used two different test rovers here on Earth in conditions designed to simulate as best as possible Spirit's predicament. However, Earth-based tests cannot exactly replicate the conditions at Troy."

Data show Spirit is straddling the edge of a 26-foot-wide crater that had been filled long ago with sulfate-bearing sands produced in a hot water or steam environment. The deposits in the crater formed distinct layers with different compositions and tints, and they are capped by a crusty soil. It is that soil that Spirit's wheels broke through. The buried crater lies mainly to Spirit's left. Engineers have plotted an escape route from Troy that heads up a mild slope away from the crater.

"We'll start by steering the wheels straight and driving, though we may have to steer the wheels to the right to counter any downhill slip to the left," said Ashley Stroupe, a JPL rover driver and Spirit extraction testing coordinator. "Straight-ahead driving is intended to get the rover's center of gravity past a rock that lies underneath Spirit. Gaining horizontal distance without losing too much vertical clearance will be a key to success. The right front wheel's inability to rotate greatly increases the challenge."

Spirit has been examining its Martian surroundings with tools on its robotic arm and its camera mast. The rover's work at Troy has augmented earlier discoveries it made indicating ancient Mars had hot springs or steam vents, possible habitats for life. If escape attempts fail, the rover's stationary location may result in new science findings.

"The soft materials churned up by Spirit's wheels have the highest sulfur content measured on Mars," said Ray Arvidson a scientist at Washington University in St. Louis and deputy principal investigator for the science payloads on Spirit and Opportunity. "We're taking advantage of its fixed location to conduct detailed measurements of these interesting materials."

Spirit and its twin rover landed on Mars in January 2004. They have explored Mars for five years, far surpassing their original 90-day mission. Opportunity currently is driving toward a large crater called Endeavor.

NASA's JPL manages the rovers for NASA's Science Mission Directorate in Washington.

Robot Firm Aims for Contract to Build Lunar Launch Pad*

By David Templeton
Pittsburgh Post-Gazette, March 4, 2009

A Pittsburgh robotics company, already competing for the Google Lunar X Prize, is making future plans for robots—hopefully its own—to build NASA's first lunar launch pad and landing site.

Astrobotic Technology Inc., an Oakland-based company led by renowned Carnegie Mellon University roboticist William "Red" Whittaker, has announced its strategy to use robots to build a moon pad at one of the moon's poles.

With technical assistance from CMU's Robotics Institute, Astrobotic presented its strategy Friday during a NASA Lunar Surface Systems conference co-sponsored by the U.S. Chamber of Commerce and its Space Enterprise Council in Washington, D.C.

The pad is scheduled for completion in 2020.

The key challenge is protecting people, their habitats and equipment from moon dust, which will be sent flying at sandblasting speeds during moon landings and takeoffs.

Astrobotic's plan calls for two robots, and possibly a third for standby, to work around the clock for almost six months to build a berm to prevent dust or regolith from traveling at high speeds in low gravity and sandblasting the outpost.

A second option calls for robots to construct a rock-lined launch pad so landings and blastoffs do not disturb dust. But that option depends on whether enough rocks exist within a reasonable distance from the moon's poles.

"For efficient cargo transfer, the landing site needs to be close to the outpost's crew quarters and laboratories," Dr. Whittaker said. "Each rocket landing and takeoff, however, will accelerate lunar grit outwards from the pad. With no atmosphere to slow it down, the dry soil would sandblast the outpost."

Robots already on the drawing board will shovel moon dirt or gather rocks into bins, then transport the payload for use in building a berm or a pad.

If NASA chooses the first option, lawn-mower-sized robotic rovers weighing

330 pounds each could build an 8.5-foot-high berm in a 160-foot semicircle in fewer than six months. That project would require the robots to move 2.6 million pounds of lunar dirt.

According to the second option, small robots would comb the lunar soil for rocks to be used to pave a durable grit-free landing pad, said John Kohut, Astrobotic's chief executive officer. "This might reduce the need to build protective berms."

Before such decisions are made, robotic scouting missions are necessary to gather more details. Researchers also must determine how much force, and how much energy, is needed to dig lunar soil.

"We need to get more data about lunar soil at the polar areas—how easy it is to scrape up and collect and how abundant rocks are at the poles," David Gump, Astrobotic president, said. "Right now, we and NASA are making some pretty gross assumptions."

Moon dust—a fine, spiky, self-cohesive dust formed by micrometeoric impacts on the moon—represents the bane of lunar exploration. Human landings already have proven that moon dust readily penetrates and could threaten operation of equipment.

"The biggest technical challenge that we will face is protecting ourselves against the very fine moon dust," Mr. Gump said. "You must seal your motors, your axles and any joints."

Astrobotic and CMU officials, with their longstanding experience in building robots for difficult terrains, are interested in producing the robots for the NASA project once more information is gathered.

"We are positioning ourselves to be the first robotic picks and shovels to build NASA's first lunar base," Mr. Gump said.

Details of the study and lunar imagery are available at www.astrobotictech.com.

Dr. Whittaker also is leading Astrobotic in developing its first lunar robot, which has been undergoing field trials for months. The company's goal is to win the $20 million Google Lunar X prize by sending a robot to the Apollo 11 landing site and transmitting high-definition video back to Earth. That project is scheduled for liftoff in December 2010.

To mark the 40th anniversary of the Apollo 11 moon walk in July, Astrobotic plans a series of news conferences to demonstrate its prototype lunar robots.

5

Lifesavers or Killing Machines? The Pros and Cons of Robots in the Military

Editor's Introduction

In 2001, as part of its Defense Authorization Act, Congress ordered the U.S. military to ramp up its use of robotic technology. Specifically, lawmakers gave the Pentagon ten years to automate one-third of its deep-strike airplanes and fifteen years to do likewise with its ground vehicles—directives intended to shift the burden of combat from humans to machines.

Similar efforts are already underway, and as the United States and its allies continue to conduct combat operations in Iraq and Afghanistan, Reaper and Predator drone flyers are being used to both gather intelligence and launch attacks. While the aircraft cruise through dangerous regions home to suspected terrorists and insurgents, their pilots work the controls from bases well out of harm's way.

Proponents of these and other automated-fighting tactics tout their potential for saving the lives of U.S. troops. If such technology had been available during the D-Day invasion of France in World War II, one might argue, robots, not soldiers, would have stormed the beaches of Normandy, and the brave men who fell to enemy fire would have returned home to their families. Indeed, they may never have left at all.

As others see it, the ends—fewer flag-draped coffins returning from overseas—don't justify the immorality and potential dangers implicit in the means. Robots, no matter how well programmed, aren't people, opponents argue. Machines lack the compassion and judgment of flesh-and-blood fighters, and their presence on the streets of Baghdad, Iraq, and other war zones may undermine efforts to win the hearts and minds of civilians. What's more, by taking the "risk" out of war, some warn, robot soldiers will make governments more cavalier about the use of force, and the world will become a more violent place.

Selections in this chapter examine both sides of the issue. In "Bug-Sized Spies," the leadoff article, James Hannah ignores the pitfalls of automated fighting and looks forward to a time when robotic insects will help the U.S. military spy on enemies. Eric Stoner, author of the second piece, "Attack of the Killer Robots," is less enthusiastic about robotic warfare. He argues that giving soldiers the ability to kill via remote control will desensitize them to violence and keep them from having to face the consequences of their actions. Stoner also posits that increased automation will prove costly, lead to more civilian casualties, and possibly shift the violence from foreign battlefields to the mainland United States, as enemies attempt to strike the command centers from which the robots are controlled.

In "Mobilizing Machines," Andrew Waldman provides a fairly balanced overview of the topic. He interviews robot expert Joe Dyer, who, in making the case for using unmanned vehicles in infantry operations, states that more than half of the U.S. Army's casualties occur during "initial contact with the enemy." "What a great job for a robot," Dyer says. On the other hand, Waldman considers what might happen if conflicts become "riskless," and those with access to robot fighters can launch attacks without fear of racking up massive body counts—at least on their own side.

The next selection, a satirical piece titled "Forget Everything Bad I Said About Robots," finds writer Scott Feschuk issuing an ironic defense of robot warfare. "So there you have it—robots with guns: safe, smart, a good idea," he observes, having run through a litany of reasons why battlefield automation is ill conceived. In "Killer Robots," Gordon S. Clark is more straightforward with his criticism of drone aircraft, wondering how anyone can stand behind a technology that, thus far in Pakistan, has been found to kill 50 civilians for every one al Qaeda leader.

In "Robots on the Battlefield," the final article, the author interviews, among others, Ellen Purdy, an enterprise director at the U.S. Pentagon, who admits that serious questions remain regarding how robots might conform to accepted laws of warfare. "There's a threshold where just because you can, doesn't mean you should," Purdy says.

Bug-Sized Spies*

U.S. Develops Tiny Flying Robots

By James Hannah
The Associated Press, November 21, 2008

If only we could be a fly on the wall when our enemies are plotting to attack us. Better yet, what if that fly could record voices, transmit video and even fire tiny weapons?

That kind of James Bond-style fantasy is actually on the drawing board. U.S. military engineers are trying to design flying robots disguised as insects that could one day spy on enemies and conduct dangerous missions without risking lives.

"The way we envision it is, there would be a bunch of these sent out in a swarm," said Greg Parker, who helps lead the research project at Wright-Patterson Air Force Base in Dayton [Ohio]. "If we know there's a possibility of bad guys in a certain building, how do we find out? We think this would fill that void."

In essence, the research seeks to miniaturize the Unmanned Aerial Vehicle drones used in Iraq and Afghanistan for surveillance and reconnaissance.

The next generation of drones, called Micro Aerial Vehicles, or MAVs, could be as tiny as bumblebees and capable of flying undetected into buildings, where they could photograph, record, and even attack insurgents and terrorists.

By identifying and assaulting adversaries more precisely, the robots would also help reduce or avoid civilian casualties, the military says.

Parker and his colleagues plan to start by developing a bird-sized robot as soon as 2015, followed by the insect-sized models by 2030.

The vehicles could be useful on battlefields where the biggest challenge is collecting reliable intelligence about enemies.

"If we could get inside the buildings and inside the rooms where their activities are unfolding, we would be able to get the kind of intelligence we need to shut them down," said Loren Thompson, a defense analyst with the Lexington Institute in Arlington, Va.

Philip Coyle, senior adviser with the Center for Defense Information in Washington D.C., said a major hurdle would be enabling the vehicles to carry the weight of cameras and microphones.

"If you make the robot so small that it's like a bumblebee and then you ask the bumblebee to carry a video camera and everything else, it may not be able to get off the ground," Coyle said.

Parker envisions the bird-sized vehicles as being able to spy on adversaries by flying into cities and perching on building ledges or power lines. The vehicles would have flappable wings as a disguise but use a separate propulsion system to fly.

"We think the flapping is more so people don't notice it," he said. "They think it's a bird."

Unlike the bird-sized vehicles, the insect-sized ones would actually use flappable wings to fly, Parker said.

He said engineers want to build a vehicle with a 1-inch wingspan, possibly made of an elastic material. The vehicle would have sensors to help avoid slamming into buildings or other objects.

Existing airborne robots are flown by a ground-based pilot, but the smaller versions would fly independently, relying on preprogrammed instructions.

Parker said the tiny vehicles should also be able to withstand bumps.

"If you look at insects, they can bounce off of walls and keep flying," he said. "You can't do that with a big airplane, but I don't see any reason we can't do that with a small one."

An Air Force video describing the vehicles said they could possibly carry chemicals or explosives for use in attacks.

Once prototypes are developed, they will be flight-tested in a new building at Wright-Patterson dubbed the "micro aviary" for Micro Air Vehicle Integration Application Research Institute.

"This type of technology is really the wave of the future," Thompson said. "More and more military research is going into things that are small, that are precise and that are extremely focused on particular types of missions or activities."

Attack of the Killer Robots*

By Eric Stoner
Boise Weekly (Idaho), August 19–25, 2009

One of the most captivating storylines in science fiction involves a nightmarish vision of the future in which autonomous killer robots turn on their creators and threaten the extinction of the human race. Hollywood blockbusters such as *Terminator* and *The Matrix* are versions of this cautionary tale, as was *R.U.R. (Rossum's Universal Robots)*, the 1920 Czech play by Karel Capek that marked the first use of the word "robot."

In May 2007, the U.S. military reached an ominous milestone in the history of warfare—one that took an eerie step toward making this fiction a reality. After more than three years of development, the U.S. Army's 3rd Infantry Division based south of Baghdad deployed armed ground robots.

Although only three of these weaponized "unmanned systems" have hit Iraq's streets, to date, *National Defense* magazine reported in September 2007 that the Army has placed an order for another 80.

A month after the robots arrived in Iraq, they received "urgent material release approval" to allow their use by soldiers in the field. The military, however, appears to be proceeding with caution.

According to a statement by Duane Gotvald, deputy project manager of the Defense Department's Robotic Systems Joint Project Office, soldiers are using the robots "for surveillance and peacekeeping/guard operations" in Iraq. By all accounts, robots have not fired their weapons in combat since their deployment more than a year and a half ago.

But it is only a matter of time before that line is crossed.

FUTURE FIGHTING FORCE?

For many in the military-industrial complex, this technological revolution could not come soon enough.

Robots' strategic impact on the battlefield, however—along with the moral and ethical implications of their use in war—have yet to be debated.

Designed by Massachusetts-based defense contractor Foster-Miller, the Special Weapons Observation Remote Direct-Action System, or SWORDS, stands 3 feet tall and rolls on two tank treads.

It is similar to the company's popular TALON bomb disposal robot—which the U.S. military has used on more than 20,000 missions since 2000—except, unlike TALON, SWORDS has a weapons platform fixed to its chassis.

Currently fitted with an M249 machine gun that fires 750 rounds per minute, the robot can accommodate other powerful weapons, including a 40 mm grenade launcher or an M202 rocket launcher.

Five cameras enable an operator to control SWORDS from up to 800 meters away with a modified laptop and two joysticks. The control unit also has a special "kill button" that turns the robot off should it malfunction. (During testing, it had the nasty habit of spinning out of control.)

Developed on a shoestring budget of about $4.5 million, SWORDS is a primitive robot that gives us but a glimpse of things to come. Future models—including several prototypes being tested by the military—promise to be more sophisticated.

Congress has been a steady backer of this budding industry, which has a long-term vision for technological transformation of the armed forces.

In 2001, the Defense Authorization Act directed the Pentagon to "aggressively develop and field" robotic systems in an effort to reach the ambitious goal of having one-third of the deep-strike aircraft unmanned within 10 years, and one-third of the ground combat vehicles unmanned within 15 years.

To make this a reality, federal funding for military robotics has skyrocketed. From fiscal year 2006 through 2012, the government will spend an estimated $1.7 billion on research for ground-based robots, according to the congressionally funded National Center for Defense Robotics. This triples what was allocated annually for such projects as recently as 2004.

The centerpiece of this roboticized fighting force of the future will be the 14 networked, manned and unmanned systems that will make up the Army's Future Combat System—should it ever get off the ground. The creation of the weapons systems is also one of the most controversial and expensive the Pentagon has ever undertaken.

In July 2006, the Defense Department's Cost Analysis Improvement Group estimated that its price tag had risen to more than $300 billion—an increase of 225 percent over the Army's original $92 billion estimate in 2003, and nearly half of President Barack Obama's proposed stimulus package.

"WAR IN A CAN"

Despite the defense world's excitement and the dramatic effect robots have on how war is fought, U.S. mainstream media coverage of SWORDS has been virtually nonexistent.

Worse, the scant attention these robots have received has often been little more than free publicity. *Time* magazine, for example, named SWORDS one of the "coolest inventions" of 2004. "Insurgents, be afraid," is how its brief puff piece began. And while most articles are not that one-sided, any skepticism is usually mentioned as a side note.

On the other hand, prior to the deployment of SWORDS, numerous arguments in their defense could regularly be found in the press. According to their proponents—generally the robot's designers or defense officials—robots will not have any of the pesky weaknesses of flesh-and-blood soldiers.

"They don't get hungry," Gordon Johnson, who headed a program on unmanned systems at the Joint Forces Command at the Pentagon, told the *New York Times* in 2005. "They're not afraid. They don't forget their orders. They don't care if the guy next to them has just been shot. Will they do a better job than humans? Yes."

Ronald Arkin, a leading roboticist at Georgia Tech, whose research is funded by the Defense Department, argues without a sense of irony that autonomous robots will be more humane than humans. Atrocities like the massacre by U.S. troops in Haditha, Iraq, would be less likely with robots, he told *The Atlanta Journal-Constitution* in November 2007, because they won't have emotions that "cloud their judgment and cause them to get angry."

Robots are also promoted as being cost-effective. On top of the annual salary and extra pay for combat duty, the government invests a great deal in recruiting, training, housing and feeding each soldier. Not to mention the costs of health care and death benefits, should a soldier be injured or killed.

By comparison, the current $245,000 price tag on SWORDS—which could drop to $115,000 per unit if they are mass-produced—is a steal.

After attending a conference on military robotics in Baltimore, journalist Steve Featherstone summed up their function in *Harper's* in February 2007: "Robots are, quite literally, an off-the-shelf war-fighting capability—war in a can."

And the most popular talking point in favor of armed robots is that they will save U.S. soldiers' lives. To drive the point home, proponents pose this rhetorical question: Would you rather have a machine get blown up in Iraq, or your son or daughter?

REMOVE FROM REALITY

At first glance, these benefits of military robots sound sensible. But they fall apart upon examination.

Armed robots will be far from cost effective. Until these machines are given greater autonomy—which is currently a distant goal—the human soldier will not be taken out of the loop. And because each operator can now handle only one robot, the number of soldiers on the Pentagon's payroll will not be slashed anytime soon. More realistically, SWORDS should best be viewed as an additional, expensive remote-controlled weapons system at the military's disposal.

A different perspective is gained when the price of the robot is compared with the low-tech, low-cost weaponry that U.S. forces face on a daily basis in Iraq.

"You don't want your defenses to be so expensive that they'll bankrupt you," said Sharon Weinberger, a reporter for *Wired*'s Danger Room blog. "If it costs us $100,000 to defeat a $500 roadside bomb, that doesn't sound like such a good strategy—as pretty as it may look on YouTube and in press releases."

The claim that robots would be more ethical than humans similarly runs contrary to both evidence and basic common sense.

Lt. Col. Dave Grossman wrote in his 1996 book *On Killing* that despite the portrayal in our popular culture of violence being easy, "There is within most men an intense resistance to killing their fellow man. A resistance so strong that, in many circumstances, soldiers on the battlefield will die before they can overcome it."

One of the most effective solutions to this quandary, the military has discovered, is to introduce distance into the equation. Studies show that the farther the would-be killer is from the victim, the easier it is to pull the trigger. Death and suffering become more sanitized—the humanity of the enemy can be more easily denied. By giving the Army and Marines the capability to kill from greater distances, armed robots will make it easier for soldiers to take life without troubling their consciences.

The Rev. G. Simon Harak, an ethicist and the director of the Marquette University Center for Peacemaking, said, "Effectively, what these remote control robots are doing is removing people farther and farther from the consequences of their actions."

Moreover, the similarity that the robots have to the life-like video games that young people grow up playing will blur reality further.

"If guys in the field already have difficulties distinguishing between civilians and combatants," Harak asked, "what about when they are looking through a video screen?"

Rather than being a cause for concern, however, Maj. Michael Pottratz at the Army's Armament Research, Development and Engineering Center in Picatinny Arsenal, N.J., wrote in an e-mail that developers are in the process of making the control unit for the SWORDS more like a "Game Boy-type controller."

It is not only possible but likely that a surge of armed robots would lead to an increase in the number of civilian casualties, not a decrease.

The supposed conversation-ender that armed robots will save U.S. lives isn't nearly as clear as it is often presented, either. "If you take a narrow view, fewer soldiers would die," Harak said, "but that would be only on the battlefield."

As happens in every war, however, those facing new technology will adapt to them.

"If those people being attacked feel helpless to strike at the robots themselves, they will try to strike at their command centers," Harak said, "which might well be back in the United States or among civilian centers. That would then displace the battlefield to manufacturing plants and research facilities at universities where such things are being invented or assembled . . . The whole notion that we can be invulnerable is just a delusion."

THE NEW MERCENARIES

Even if gun-toting robots could reduce U.S. casualties, other dangerous consequences of their use are overlooked.

Frida Berrigan, a senior program associate at the New America Foundation's Arms and Security Initiative, argues that similar to the tens of thousands of unaccountable private security contractors in Iraq, robots will help those in power "get around having a draft, higher casualty figures and a real political debate about how we want to be using our military force."

In effect, by reducing the political capital at stake, robots will make it far easier for governments to start wars in the first place.

Since the rising U.S. death toll appears to be the primary factor that has turned Americans against the war—rather than its devastating economic costs or the far greater suffering of the Iraqi people—armed robots could also slow the speed with which future wars are brought to an end.

When Arizona Republican Sen. John McCain infamously remarked that he would be fine with staying in Iraq for 100 years, few noted that he qualified that statement by saying, "as long as Americans are not being injured or harmed or wounded or killed."

Robot soldiers will be similar to mercenaries in at least one more respect. They both serve to further erode the state's longstanding monopoly on the use of force.

"If war no longer requires people, and robots are able to conduct war or acts of war on a large scale, then governments, will no longer be needed to conduct war," Col. Thomas Cowan Jr. wrote in a March 2007 paper for the U.S. Army War College. "Non-state actors with plenty of money, access to the technology and a few controllers will be able to take on an entire nation, particularly one which is not as technologically advanced."

This may not be farfetched.

In December 2007, *Fortune* magazine told the story of Adam Gettings, "a 25-year-old self-taught engineer," who started a company in Silicon Valley called Robotex. Within six months, the company built an armed robot similar to the SWORDS except that it costs a mere $30,000 to $50,000. And these costs will drop.

As this happens, and as the lethal technology involved becomes more accessible, Noel Sharkey, a professor of Artificial Intelligence and Robotics at the University of Sheffield in the United Kingdom, warns that it will be only a matter of time before extremist groups or terrorists develop and use robots.

Evidence now suggests that using armed robots to combat insurgencies would be counterproductive from a military perspective.

One study, published in the journal *International Organization* in June 2008, by Jason Lyall, an associate professor of international relations at Princeton, and Lt. Col. Isaiah Wilson III, who was the chief war planner for the 101st Airborne Division in Iraq and who currently teaches at West Point, looks at 285 insurgencies dating back to 1800.

After analyzing the cases, Lyall and Wilson concluded that the more mechanized a military is, the lower its probability of success.

"All counterinsurgent forces must solve a basic problem: How do you identify the insurgents hiding among noncombatant populations and deal with them in a selective, discriminate fashion?" Lyall wrote in an e-mail.

To gain such knowledge, troops must cultivate relationships with the local population. This requires cultural awareness, language skills and, importantly, a willingness to share at least some of the same risks as the local population.

The *Counterinsurgency Field Manual*, which was released in December 2006 and co-authored by Gen. David Petraeus, would seem to agree.

"Ultimate success in COIN [counterinsurgency] is gained by protecting the populace, not the COIN force," the manual states. "If military forces remain in their compounds, they lose touch with the people, appear to be running scared, and cede the initiative to the insurgents."

Mechanized militaries, however, put greater emphasis on protecting their own soldiers. Consequently, Lyall and Wilson argued in their study that such forces lack the information necessary to use force discriminately, and therefore, "often inadvertently fuel, rather than suppress, insurgencies."

Given such findings, deploying armed robots in greater numbers in Iraq or Afghanistan would likely only inflame resistance to the occupation, and, in turn, lead to greater carnage.

To understand this point, put yourself in the shoes of an Iraqi or Afghani. How could seeing a robot with a machine gun rumble down your street or point its weapon at your child elicit any reaction other than one of terror or extreme anger? What would you do under such circumstances? Who would not resist? And how would you know that someone is controlling the robot?

For all the Iraqis know, SWORDS is the autonomous killer of science fiction—American-made, of course.

The hope that killer robots will lower U.S. casualties may excite military officials and a war-weary public, but the grave moral and ethical implications—not to mention the dubious strategic impact—associated with their use should give pause to those in search of a quick technological fix to our woes.

By distancing soldiers from the horrors of war and making it easier for politicians to resort to military force, armed robots will likely give birth to a far more dangerous world.

Mobilizing Machines*

Robots and Drones Are Beginning to Change the Nature of Combat

By Andrew Waldman
National Guard, July 2009

Picture this: A fleet of the U.S. Navy's most advanced warships arrives off the shores of a rogue nation. Its mission is to provide intelligence and an initial landing force to clear the way for a large-scale invasion.

Unlike their 20th-century predecessors, commanders deploy no amphibious troop transports. There are no combat engineers on the shores blowing through defensive positions. And there are no fighter planes flying low overhead.

Instead, the ships launch unmanned, submersible robots that stealthily seek out and clear anti-ship mines and enemy submarines. Then, larger unmanned boats are launched carrying platoons of ground-based fighting robots that hit the shores, climb cliffs, clear paths and destroy enemy positions. Above them, unmanned aerial vehicles (UAVs) are providing intelligence and close-air support.

The robots fight for hours. Some are lost, but their operators, including, most likely, National Guardsmen, are far from the danger zone, sitting safely in operations centers in Nevada and North Dakota. Only when the shores are finally secured and the smoke has cleared do fleet commanders notify soldiers—the human type—that they can disembark to finish the fight.

This scenario is obviously fictional. But military robotics experts like Joe Dyer, who leads the government and industrial robots division at the firm iRobot, says it could be the not-so-distant future of war fighting for the U.S. military.

Robotics has been one of the most rapidly developing areas in the military over the last decade. There are thousands of missions using unmanned aerial and ground vehicles in Iraq and Afghanistan, and the use of these mechanical warriors is not going to slow down.

* Courtesy of the U.S. Department of Defense.

Guard units are already involved in operations that use robotics, and there no doubt will be more and more robotic weapons systems in the Guard in the future.

Four Air Guard units fly the Predator, a UAV being used heavily in Afghanistan. A fifth unit will be added in the first quarter of fiscal year 2010. And another Air Guard unit will soon fly the Reaper, a larger and more lethal UAV.

Robotics also has a major role in explosive ordnance disposal. Technicians like Sgt. 1st Class Stuart Stevens, a member of the North Carolina Army Guard's 430th Ordnance Company (EOD), know the advantages of robotics.

Stevens, who serves as his unit's acting first sergeant, readiness NCO and maintenance technician, has been an EOD tech since 1989, and remembers the first robots that made their way to the job specialty years ago.

In the Army, most, if not all, EOD technicians use robots, which, Stevens says, keep soldiers out of harm's way.

The 430th uses three different robots, including iRobot's PackBot, the remote ordnance neutralization system (RONS) and the Foster-Miller TALON, which has been used for EOD missions since 2000.

Each has a specific use, but the main idea is the same: The robot can approach and inspect any type of unexploded ordnance from an old mine to a live improvised explosive device.

And they look similar, too. EOD robots are, for the most part, small, tracked vehicles that sit low to the ground. An iRobot PackBot with an EOD kit looks like a cross between a broomstick and a radio-controlled tank, with its extendable bomb removal arm and a camera swiveling around to give operators a live view of the danger zone.

Some robots can move at a pace of 5 mph for up to six hours and carry gear with them. Many are equipped with microphones and speakers so operators can communicate with any people the robot might encounter.

And unlike their human counterparts, they don't get tired, scared or hungry. As long as there is an alert operator at the other end, the robot can keep working.

Some can even start the process of disarming the dangerous ordnance before soldiers have to approach it.

In a unit like the 430th, the robot has become a vital member of the team. EOD companies have an option with robots that did not exist 20 years ago when soldiers like Stevens started their careers.

"It gives us a choice," he says. "If someone calls in something, in the old days, I'd have to get out of my jeep and walk down and look at it. Now, I can stay inside my vehicle, put the robot out [and] the robot goes down to look at it."

Robots, however, are not perfect. All machines—even robots—can break.

"Robotics has made the field a lot safer," Stevens says, but adds, "EOD is never a safe job. A lot of things can go wrong and the robots don't always work. If [soldiers] don't have a robot, they still have to do it manually."

Though the most prominent use of ground-based robotic unmanned vehicles has been in the EOD field, it isn't the only area pursued by manufacturers of the technology.

According to Dyer, who retired from the U.S. Navy as a vice admiral commanding Naval Air Systems Command, other vehicles that have infantry uses are on the way, including, most notably, the small unmanned ground vehicle (SUGV), which is part of the Army's Future Combat System.

Defense Secretary Robert M. Gates has proposed cutting some portions of the FCS program, but not the unmanned ground vehicles. Dyer says the reason for the continued interest in the unmanned infantry systems is that they assume the risk during the first moments of combat.

"Robots are becoming a tool for the infantry. One of the points we hear from the Army's leadership is 52 percent of the Army's casualties come from initial contact with the enemy," says Dyer. "What a great job for a robot."

During testing of the SUGV Dyer says, the soldiers testing the device at Fort Bliss, Texas, were originally skeptical.

"When we delivered (the SUGV), they said, 'We don't need no stinkin' robots,'" he says. "Two weeks later, we couldn't get them to let go of the robots so we could do more developments with them."

Dyer says this type of robot could turn out to be an "infantryman's Swiss army knife," as it could be used to provide reconnaissance and improve situational awareness.

Some robotic systems are going beyond those tasks.

The Foster-Miller TALON, which Stevens' EOD unit uses, has been modified through an Army weapons program to be a fully capable combat robot. The special weapons observation reconnaissance detection system (SWORDS) can carry a variety of mounted weapons, including an M-249 or M-16. It also has a full array of imaging equipment and is as accurate as a sniper, even as it calmly acquires targets in the middle of a fire fight.

Although many of the robots in the military are used solely for overseas operations, there are a number of domestic uses for them. According to Gary Edelblute, the National Guard Bureau's airborne intelligence, surveillance and reconnaissance and unmanned aircraft systems subject-matter expert, UAVs were used after Hurricane Katrina. The North Dakota Guard also employed them to assist its response to spring flooding.

But because most unmanned aerial and ground vehicles are overseas, the Guard has not done extensive domestic training with them. But Stevens says there are definitely uses for the robots stateside, like sampling chemical hazards from a distance or delivering phones or information into buildings during hostage situations.

For soldiers like Stevens, the robot is simply the best way to accomplish the mission with minimal risk. And robotic technology cunently in the field hasn't yet been utilized to its full potential.

Stevens cites the RONS, a 700-pound robot, as an example. The robot is capable of a variety of tasks, from moving an EOD to rescuing an injured soldier. But because of its bulky size, it is hard to move and, thus, not utilized as often as smaller, more mobile robots. But give soldiers time and they'll figure it out.

"They are always thinking about ways to use things," Stevens says about EOD techs who operate robots. "The limitations are only those that the soldiers place on themselves."

With the advent of unmanned combat vehicles come difficult decisions about the definition of conflict and the future of warfare in the world. According to PW Singer, military technology expert and author of *Wired For War*, the very definition of war could be completely changed by robots, and military leadership is still grappling with how to use the technology in the field.

"At a [high] level, how does this affect things like 'going to war,'" Singer says.

"It's meant the same thing for 5,000 years for everyone. At its most basic level, it meant going to a place [to fight] and you might not come back."

Now, UAV pilots, like those in the North Dakota Air Guard's 119th Wing, can sit in a cubicle in Fargo flying missions over Afghanistan for eight hours and return to their homes in time to eat dinner with their families. That creates a new set of challenges at all levels of combat, Singer says.

"This is a break point in history. They are at war, but they are at home," says Singer. "They are at war with no risk other than carpal tunnel."

Singer says such a disconnection from the physical fight could mean that countries with access to robotic technology could consider armed conflicts "riskless" and deploy robots without fear of casualties.

At this point, the field of robotics is in its infancy, with only a small portion of the potential power of the technology tapped for use in the real world. Singer likens this point in robotic history to where combustible engine technology was in 1908 or where computers were 25 years ago.

Robots will probably become as ubiquitous as computers, Singer says. There are already many robots in the world, doing everything from building cars to vacuuming carpets in the home, as well as sniffing out bombs. At some point, he says, society won't even notice them.

"It's like your car," he says. "There are hundreds of computers in your car, but you don't call it a 'computer car.'"

Forget Everything Bad I Said About Robots*

I Never Grasped that We Are the Flesh-Based Problem to Which They Are the Solution

By Scott Feschuk
Maclean's, March 23–30, 2009

It's been a while since I raised the potential threat posed by robots. In fact, it's been so long that some readers have emailed to accuse me of having been bought off and silenced by the menacing robo-industrial complex. Let me assure you: nothing, with the exception of a Conservative TV commercial depicting Stephen Harper as empathetic, could be further from the truth.

But my thinking has definitely evolved. A year ago, I described the many horrors of the forthcoming robocalypse and how—thanks to advances in robotics—all humanity is destined to lead lives that are much more leisurely and, come the blood-soaked dawn of the robot revolution, much more over.

I stated my belief that armed robots would ultimately rise up against their creators, using their advanced programming and pinchy claws to purge the earth of the vile human stain. But boy, was I wrong. Robots are great! And I'm not saying that because I'm currently being held against my will by my Roomba.

My doubt about the survival of our species was assuaged by a sunny new report entitled "Autonomous Military Robotics," which was written for U.S. military planners. The document envisions a utopian future in which wars are waged primarily by machines. The worst thing that could happen to you as a human during such a conflict? Your blender might get drafted. And even then you'd stand a good chance of being awarded the Victoria Cross for valour in the face of smoothielessness.

"Imagine the face of warfare with autonomous robotics," the report begins, almost gleefully. "Instead of our soldiers returning home in flag-draped caskets to heartbroken families, autonomous robots can replace the human soldier in an increasing range of dangerous missions." The authors, a trio of California university researchers, make the case that we are moving ever closer to the glorious day when robots will develop a sense of identity and be able to think and reason for themselves, just like 33 per cent of the Jonas Brothers: "These robots would be 'smart' enough to make decisions that only humans now can." (In fact, encouraging new research suggests some toasters are already capable of debating the ladies on *The View*.)

What really won me over in this report is its down-on-humanity tone. I'd never fully grasped that we are the flesh-based problem to which unstoppable robotic killing machines are the gleaming solution. The researchers seem to delight in noting that "robots have a distinct advantage over the limited and fallible cognitive capabilities that we Homo sapiens have." For instance, if robots noticed that an endless series of movies were being made about robots turning evil and taking over the world, robots would probably be smart enough not to build robots like that. But not us!

Nodding to skeptics, the report's authors do acknowledge that the process of developing and deploying heavily armed, autonomous soldier-robots won't be without its "growing pains." In fact, they even use that strangely colloquial expression—growing pains—and in so doing essentially equate being hunted down and brutally dismembered by a haywire robot to the experience of watching the 1980s sitcom starring Alan Thicke, Kirk Cameron and . . . actually, that seems like a pretty fair comparison.

Wisely, the report makes only scant mention of the "semi-autonomous robotic cannon" in South Africa that shot 23 "friendly" soldiers (apparently, those who survived were noticeably less "friendly" to the cannon afterwards), or the epidemic crashing of drone aircraft around the world, or the incident from last April in which several U.S. units of Iraq-deployed Talon Swords—mobile robots armed with machine guns—abruptly trained their guns on American soldiers. Sure, these chronic screw-ups may well be harbingers of the grave and fatal consequences that will ultimately be exacted by our hubris—but then again, there's a remote chance they theoretically might possibly not be. So let's go with that.

What's important is that any anxiety being felt by human military personnel in Iraq be downplayed. I mean, some of these U.S. soldiers act as though they've never been commanded to fight a well-armed insurgency while simultaneously fleeing their own lethally unhinged robotic death tools. Come on! It was all covered in the army's basic-training manual, under the section entitled "How Did All These Bullets Get In My Torso?"

And hey—if a robot does shoot you, there's a chance you could be saved by . . . a robot. A California technology firm is currently building a three-armed robot that moves on treads and is programmed to replace medics in providing urgent medical intervention on the battlefield. A spokesman claims: "It could relieve immediate life-threatening injuries, or stop bleeding temporarily." Sometimes it might even do these things to a soldier it didn't first injure by running him over.

So there you have it—robots with guns: safe, smart, a good idea. You've got the U.S. military's word on it. And when have they ever been wrong about anything?

Killer Robots*

By Gordon S. Clark
Sojourners, August 2009

When I reflect on the drone aircraft now being used by the U.S. military to kill suspected "extremists"—and large numbers of innocent bystanders—in Afghanistan and Pakistan, an episode of the original *Star Trek* comes to mind. In it, Captain Kirk accidently transports into a parallel universe run on fear and violence, where the evil "mirror" Kirk maintains his hold on power with a device that allows him to spy on anyone in the ship and to assassinate them with the press of a button.

Sound familiar?

Drone aircraft are pilotless planes operated by remote control, often from thousands of miles away. In the last two years the drones, which are equipped with both cameras and weapons, have been increasingly used to launch attacks rather than gather intelligence—a fact that has quietly and without significant protest slipped into the narrative of the U.S. war in Afghanistan and Pakistan.

Yet there is every reason to oppose their use and existence. Consider: Even worse than our overall military operations in Afghanistan, which kill significant numbers of civilians, these pilotless drones are by their nature incapable of distinguishing between combatants and civilians. Drones simply launch missiles into buildings or compounds, killing whomever happens to be there. David Kilcullen, a former adviser to Gen. David Petraeus, testified to Congress earlier this year that 14 al Qaeda leaders have been killed in Pakistan by drone attacks since 2006—along with 700 civilians. That's an astounding ratio of 50 innocent victims killed for each targeted individual!

And, because those targeted are only suspected extremists, drone attacks fundamentally subvert our core legal value of "innocent until proven guilty." The attacks are, as Catholic anti-war activist Kathy Kelly notes, nothing more than extrajudicial executions, turning mere suspicion into an automatic death sentence.

Moreover, there has never been a declaration to extend the Afghan war to Pakistan, which is where the drones are primarily used. Drone attacks are considered "covert" and are not officially discussed by President Obama or Congress—an absurd artifice that may help suppress domestic criticism, but does nothing to diminish the illegality or immorality of attacking a country with which we are not at war.

Finally, the drone attacks are, by all accounts, solidifying Pakistani public outrage against the U.S. This will inevitably drive more into the ranks of the militants, perpetuating and increasing the violence.

At the end of the *Star Trek* episode, the good Kirk tries to persuade the evil "mirror" Spock of the wisdom of cooperation over war and violence—yet gives him the assassination device as a way to make that change. President Obama seems to be following the same script. Defense Secretary Robert Gates announced in April that his proposed budget will call for "a major increase in unmanned aircraft," specifically the Predator drones now being used for attacks. In yet another ominous turn, the Pakistani government is now asking for its own drones.

How is it possible for us to create peace in this shattered region or rebuild our nation's image and role in the world while increasingly relying on an assassination device that kills more innocents than enemies and makes a mockery of our laws and values? Unlike a TV show that ends in an hour, our actions will have repercussions we will have to live with for years to come.

Robots on Battlefield*

Townsville Bulletin (Australia), September 18, 2009

Going off to war has always meant risking your life, but a wave of robotic weaponry may be changing that centuries-old truth.

The "pilots" who fly US armed drones over Afghanistan, Iraq and Pakistan sit with a joystick thousands of kilometres away, able to pull the trigger without being exposed to danger.

Other robots under development could soon ferry supplies on dangerous routes and fire at enemy tanks.

The explosion in unmanned vehicles offers the seductive possibility of a country waging war without having to put its own soldiers or civilians in the line of fire.

But analysts say the technology raises a host of ethical and legal questions, while political and military leaders have yet to fully grasp its implications.

"What's the effect on our politics? To be able to carry out operations with less human cost makes great sense. It is a great thing, you save lives," said Peter Singer, author of *Wired for War*.

"On the other hand, it may make you more cavalier about the use of force."

Commanders see unmanned vehicles as crucial to gaining the edge in combat and saving soldiers' lives, freeing up troops from what the military calls "dull, dirty and dangerous" tasks.

Cruise missiles and air strikes have already made war a more remote event for the American public.

Now, robots could offer the tantalising scenario of "pain-free" military action, said Lawrence Korb, a former US assistant secretary of defence.

"That raises the whole larger question—does it make it too easy to go to war, not just here or any place else?" he said.

Robotic technology is taking armies into uncharted territory where tens of thousands of sophisticated robots could eventually be deployed, including unmanned vehicles possibly designed to automatically open fire.

US officials insist a human will always be "in the loop" when it comes to pulling the trigger, but analysts warn that supervising robotic systems could become complicated as the technology progresses.

Military research is already moving toward more autonomous robots that will require less and less guidance.

The trend is illustrated by the US Air Force's plans to have a single human operator eventually supervise three drones at once instead of one aircraft.

Even if humans can still veto the use of force, the reality of numerous robots in combat producing a stream of information and requiring split-second decisions could prove daunting.

Future robotic weapons "will be too fast, too small, too numerous and will create an environment too complex for humans to direct," retired Army colonel Thomas Adams is quoted as saying in *Wired for War*.

Innovations with robots "are rapidly taking us to a place where we may not want to go, but probably are unable to avoid," he said.

Experience has shown humans are sometimes reluctant to override computerised weapons, placing more faith in the machine than their own judgment, according to Singer.

The military is still trying to figure out how an armed robot on the ground should be designed and operated to conform to the law of armed conflict, said Ellen Purdy, the Pentagon's enterprise director of joint ground robotics.

"Nobody has answered that question yet," Purdy said. "There's a threshold where just because you can, doesn't mean you should."

6

Thinking, Feeling Robots: The Dream (or Nightmare) of Artificial Intelligence

Editor's Introduction

It's an idea as frightening as it is popular: One day, machines will surpass humans in intelligence and take over the world. Countless science fiction books and films hinge on precisely this premise, and while the notion of global robot conquest may seem farfetched, the fact remains that machines are getting smarter all the time. Consider Big Blue, the computer that, in 1997, bested chess champion Gary Kasparov in a six-match contest. If artificial minds can master the intricacies of chess, the alarmist argument goes, what's to stop them from plotting and carrying out a successful uprising?

At least for now, those fearful of such scenarios need not begin stockpiling arms and canned goods. Sure, Big Blue defeated Kasparov, but it did so in decidedly unintelligent fashion. Instead of using intellect and strategy, as a human player would, the computer relied largely on "brute force," using its staggering data-processing abilities to select each move from millions of choices. The machine wasn't conscious of what it was doing, and when it was declared the victor, it didn't have the emotional capacity to feel good about its win. (This is lucky for Kasparov, who was spared having to watch the computer gloat.)

It turns out that programming robots to think and feel in ways comparable to humans is extremely difficult, perhaps even impossible. While some of today's advanced robots can answer questions, participate in simple conversations, and recognize facial expressions, they're more responding to programming than expressing any learned intelligence or self-awareness. "Intelligence is not only what you know," bioengineer Giulio Sandini tells Abigail Tucker in one of the articles included in this chapter. "Intelligence is acquiring information, a dynamic process." Thus far, no robot has mastered that process, though as the selections in this chapter point out, some scientists continue to chase the dream of "artificial intelligence," or AI, the advent of which would transform the world in untold ways.

In "AI Comes of Age," the first article in this chapter, Gary Anthes provides a quick history of efforts to build thinking, feeling robots. He also introduces the concept of "machine learning," the process by which manmade minds use incoming information to grow progressively smarter and more adept at performing certain tasks, such as telling smiles from frowns or predicting what books or CDs will appeal to specific Amazon.com shoppers. Anthes also touches on how

AI researchers are shifting their focus from computer science to biology and using new knowledge of the human brain to inform their work.

The author of the second selection, "New Generation Humanoid Robots Raise Ethical Dilemmas: Scientist," discusses the work of Mary-Anne Williams, head of the innovation and research laboratory at the University of Technology in Sydney, Australia. Williams has developed two state-of-the-art humanoid robots, and while they look and act like people, they lack "the ability to do common sense reasoning and build models of other people's minds which is crucial for things like communication," the professor explains. Toward the end of the selection, the writer ponders whether truly human-like robots would be entitled to the same rights and legal protections as flesh-and-blood people.

In "Robot Babies," Abigail Tucker highlights scientist Javier Movellan's Project One experiment, an effort to build a machine with the same learning capabilities as a one-year-old child. Movellan hopes the initiative will lead to advances not only in robotics, but also our understanding of childhood development. As detailed in the next piece, "Robots May Soon Have Tails, Whiskers," European researchers are hard at work on another type of learning robot—an artificial rat whose brain is designed to function like a real rodent's. "We want to make robots that are able to look after themselves and depend on humans as little as possible," scientist Agnes Guillot says.

"Dreaming of Electric Sheep" finds author Steven Johnson musing on the role emotion plays in AI. Johnson argues that while truly intelligent robots would possess emotions, they would be completely unlike our own. "The sentient machines won't act like little boys or obedient butlers after all, but they'll be even more unnerving for their radical difference," he writes, adding that the Internet might prove a useful training ground for robot minds to learn about being human.

In the final article, "Forecasts for Artificial Intelligence," Bohumir Stedron looks ahead several decades and predicts where the field of AI research is heading. By the 2020s, "The Age of AI Self Reliance," he speculates, lawmakers will have extended to some robots the same rights guaranteed humans.

AI Comes of Age*

By Gary Anthes
Computerworld, January 26, 2009

"Stair. Please fetch the stapler from the lab," says the man seated at a conference room table. The Stanford Artificial Intelligence Robot, standing nearby, replies in a nasal monotone, "I will get the stapler for you."

Stair pivots and wheels into the adjacent lab, avoiding a number of obstacles on the way. Its stereoscopic camera eyes swivel back and forth, taking in the contents of the room. It seems to think for a moment, then approaches a table for a closer look at an oblong metallic object. Its articulated arm reaches out, swivels here and there, and then gently picks up the stapler with long, rubber-clad fingers. It heads back to the conference room.

"Here is your stapler," says Stair, handing it to the man. "Have a nice day."

These are indeed nice days for artificial intelligence researchers. While Stair's performance might not seem much better than that of a dog fetching the newspaper, it's a technological tour de force unimaginable just a few years ago. Indeed, Stair represents a new wave of AI, one that integrates learning, vision, navigation, manipulation, planning, reasoning, speech and natural-language processing.

It also marks a transition of AI from narrow, carefully defined domains to real-world situations in which systems learn to deal with complex data and adapt to uncertainty.

AI has more or less followed the "hype cycle" popularized by Gartner Inc.: Technologies perk along in the shadows for a few years, then burst on the scene in a blaze of hype. Then they fall into disrepute when they fail to deliver on extravagant promises, until they eventually rise to a level of solid accomplishment and acceptance.

AI has its roots in the late 1950s but came to prominence in the "expert systems" of the 1980s. In those systems, experts—chess champions, for example—were interviewed, and their rules of logic were hard-coded in software: If Condition A occurs, then do X. But if Condition B occurs, then do Y. While they worked reasonably well for specialized tasks such as playing chess, they were "fragile," says Eric Horvitz, an AI researcher at Microsoft Research.

"They focused on capturing chunks of human knowledge, and then the idea was to assemble those chunks into reasoning systems that would have the expertise of people," Horvitz says. But they couldn't "scale," or adapt, to conditions that had not explicitly been anticipated by programmers.

Today, AI systems can perform useful work in "a very large and complex world," Horvitz says. "Because these small [software] agents don't have a complete representation of the world, they are uncertain about their actions. So they learn to understand the probabilities of various things happening, they learn the preferences [of users] and costs of outcomes and, perhaps most important, they are becoming self-aware."

These abilities derive from something called machine learning, which is at the heart of many modern AI applications. In essence, a programmer starts with a crude model of the problem he's trying to solve but builds in the ability for the software to adapt and improve with experience. Speech recognition software gets better as it learns the nuances of your voice, for example, and over time Amazon.com more accurately predicts your preferences as you shop online.

IT'S ALL ABOUT THE DATA

Machine learning is enabled by clever algorithms, of course, but what has driven it to prominence in recent years is the availability of huge amounts of data, both from the Internet and, more recently, from a proliferation of physical sensors. Carlos Guestrin, an assistant professor of computer science and machine learning at Carnegie Mellon University, combines sensors, machine learning and optimization to make sense of large amounts of complex data. For example, he says, scientists at the University of California, Los Angeles, put sensors on robotic boats to detect and analyze destructive algae blooms in waterways. AI algorithms learned to predict the location and growth of the algae. Similarly, researchers at Carnegie Mellon put sensors in a local water-distribution system to detect and predict the spread of contaminants. In both cases, machine learning enabled better predictions over time, while optimization algorithms identified the best sites for the expensive sensors.

Guestrin is also working on a system that can search a huge number of blogs and identify those few that should be read by a given user every day based on that user's browsing history and preferences. He says it may sound completely different from the task of predicting the spread of contaminants via sensors, but it's not.

"Contaminants spreading through the water distribution system are basically like stories spreading through the Web," he says. "We are able to use the same kind of modeling ideas and algorithms to solve both problems."

Guestrin says the importance of AI-enabled tools like the blog filter may take on importance far beyond their ability to save us a few minutes a day. "We are making decisions about our lives—who we [. . .] elect, and what issues we find important—based on very limited information. We don't have time to make the kind of analyses that we need to make informed decisions. As the amount of information increases, our ability to make good decisions may actually decrease. Machine learning and AI can help."

Microsoft Research has combined sensors, machine learning and analysis of human behavior in a road traffic prediction model. Predicting traffic bottlenecks would seem to be an obvious and not very difficult application of sensors and computer forecasting. But MSR realized that most drivers hardly need to be warned that the interstate heading out of town will be jammed at 5 P.M. on Monday. What they really need to know is where and when anomalies, or "surprises," are occurring and, perhaps more important, where they will occur.

So MSR built a "surprise forecasting" model that learns from traffic history to predict surprises 30 minutes in advance based on actual traffic flows captured by sensors. In tests, it has been able to predict about 50% of the surprises on roads in the Seattle area, and it is in use now by several thousand drivers who receive alerts on their Windows Mobile devices.

Few organizations need to make sense of as much data as search engine companies do. For example, if a user searches Google for "toy car" and then clicks on a Wal-Mart ad that appears at the top of the results, what's that worth to Wal-Mart, and how much should Google charge for that click? The answers lie in an AI specialty that employs "digital trading agents," which companies like Wal-Mart and Google use in automated online auctions.

Michael Wellman, a University of Michigan professor and an expert in these markets, explains: "There are millions of keywords, and one advertiser may be interested in hundreds or thousands of them. They have to monitor the prices of the keywords and decide how to allocate their budget, and it's too hard for Google or Yahoo to figure out what a certain keyword is worth. They let the market decide that through an auction process."

When the "toy car" query is submitted, in a fraction of a second Google looks up which advertisers are interested in those keywords, then looks at their bids and decides whose ads to display and where to put them on the page.

"The problem I'm especially interested in," Wellman says, "is how should an advertiser decide which keywords to bid on, how much to bid and how to learn over time—based on how effective their ads are—how much competition there is for each keyword."

The best of these models also incorporate mechanisms for predicting prices in the face of uncertainty, he says. Clearly, none of the parties can hope to optimize the financial result from each transaction, but they can improve their returns over time by applying machine learning to real-time pricing and bidding.

BRAINY STUDIES

One might expect AI research to start with studies of how the human brain works. But most AI advances have come from computer science, not biology or cognitive science. These fields have sometimes shared ideas, but their collaboration has been at best a "loose coupling," says Tom Mitchell, a computer scientist and head of the Machine Learning Department at Carnegie Mellon University. "Most of the progress in AI has come from good engineering ideas, not because we see how the brain does it and then mimic that."

But now that's changing, he says. "Suddenly, we have ways of observing what the brain is really doing, via brain imaging methods like functional MRI. It's a way to look into the brain while you are thinking and see, once a second, a movie of your brain's activity with a resolution of 1mm."

So, cognitive science and computer science are now poised to share ideas as they never could before, he says. For example, certain AI algorithms send a robot a little reward signal when it does the right thing and a penalty signal when it makes a mistake. Over time, these have a cumulative effect, and the robot learns and improves. Mitchell says researchers have found with functional MRIs that regions of the brain behave exactly as predicted by these "reinforcement learning" algorithms. "AI is actually helping us develop models for understanding what might be happening in our brains," he says.

Mitchell and his colleagues have been examining the neural activity revealed by brain imaging to decipher how the brain represents knowledge. To train their computer model, they presented human subjects with a list of 60 nouns—such as telephone, house, tomato and arm—and observed the brain images that each produced. Then, using a trillion-word text database from Google, they determined the verbs that tend to appear with the 60 base words—ring with telephone, for example—and they weighted those words according to the frequency of both occurring.

The resulting model was able to accurately predict the brain image that would result from a word for which no image had ever before been observed. Oversimplifying, the model would, for example, predict that the noun airplane would produce a brain image more like that for train than for tomato.

"We were interested in how the brain represents ideas," Mitchell says, "and this experiment could shed light on a question AI has had a lot of trouble with: What is a good, general-purpose representation of knowledge?" There may be other lessons as well. Noting that the brain is also capable of forgetting, he asks, "Is that a feature or a bug?"

Andrew Ng, an assistant professor of computer science at Stanford University, led the development of the multitalented Stair. He says the robot is evidence that many previously separate fields within AI are now mature enough to be integrated "to fulfill the grand AI dream."

And just what is that dream? "Early on, there were famous predictions that within a relatively short time computers would be as intelligent as people," he says. "We still hope that some time in the future computers will be as intelligent as we are, but it's not a problem we'll solve in 10 years. It may take over 100 years."

AI: WHAT'S DIFFERENT NOW?

- **Ubiquitous computing and more-powerful computers**
- **Huge amounts of data from the Internet and physical sensors**
- **Algorithms that learn and improve over time**
- **Software that's able to deal with uncertainty, incompleteness and surprises**
- **Software agents that can weigh costs and benefits**
- **Integration of separate fields such as speech, vision, robotics, sensors and machine learning**

AI ON WALL STREET

Among the villains in the current financial debacle are the Wall Street "quants"—the computer scientists and mathematicians who wrote AI models for trading optimization and risk analysis. But Michael Wellman, a professor of computer science and engineering at the University of Michigan, says it's too early to say if any of the models failed to prevent the meltdown or even contributed to it.

"I think automation of trading and risk analysis is part of the solution, not part of the problem," says Wellman, who specializes in AI applications in markets. "For example, a big problem is a lack of transparency—companies not even understanding the assets they have. You'd hope that if the [investment] contracts were more machine-readable and -analyzable, companies would do much better risk analyses and would better understand what their positions are."

Stair, the Stanford Artificial Intelligence Robot, features a new generation of AI software that integrates learning, vision, planning, reasoning, speech and more. Microsoft Research's traffic forecasting system takes information captured by sensors and combines that with what it has learned from past traffic patterns to predict surprise backups 30 minutes in advance.

New Generation Humanoid Robots Raise Ethical Dilemmas: Scientist*

Asia Pulse, December 4, 2007

Angelina is "beautiful" and "very striking," according to Professor Mary-Anne Williams.

At just 45cm tall she is a miniature mechanical version of her famous Hollywood namesake.

Angelina is one of two advanced robots developed by Professor Williams, who heads up the innovation and research laboratory at Australia's University of Technology in Sydney.

She is also an example of a new generation of humanoid robots which combine human-like movement and artificial intelligence (AI).

As these robots become increasingly sophisticated, questions are inevitably arising about just how "human" these humanoids can really become.

There's no question that robots are getting better at moving and acting like us.

This was demonstrated by the recent visit to Australia of the talking, dancing, soccer-playing ASIMO—an Advanced Step in Innovative Mobility robot created by car maker Honda.

Standing at 1.3 metres, he is designed to help out in the home or office environment.

Sophisticated robots have been toiling away in industry for decades, but humanoids that actually walk and talk like us are relatively new.

"Being a robot in the shape of a human is useful for inhabiting human spaces," says Prof Williams, who wasn't involved in the ASIMO project but describes him as "an extraordinary robot."

"In a typical house, for example, if you're in a human form you don't have trouble moving about.

"But if you have wheels you're going to have trouble with stairs, going around corners and stepping over things."

Honda's first robot, EO, took 20 seconds to take a single step in 1986.

Today, ASIMO can run at 6km/h with both feet leaving the ground at once—just like a human.

Prof Williams says AI is where the real robotic frontier lies.

"We can build robots that look like us and move like us and ASIMO in motion is extraordinary," she says.

"But what he lacks is the ability to do common sense reasoning and build models of other people's minds which is crucial for things like communication."

Prof Williams believes genuine AI "will happen" and it could happen soon if there's a scientific breakthrough.

"It could happen tomorrow, it could happen in 50 years, it could happen in 100 years," she says.

"People and animals are just chemical bags, chemical systems, so there's no technical reason why we couldn't have robots that truly have AI."

Prof Williams believes a unique form of robotic emotion could even evolve one day.

"You could argue some robots can mimic (emotions) already," she says.

"But because a robot will experience the world differently to us it will be quite an effort for the robot to imagine how we feel about something."

This could pose problems down the track.

Humans generally anticipate how another person might feel about something by thinking about how it would affect them.

People who don't have the ability to empathise can become psychopaths.

"I think there is a danger of producing robots that are psychopathic," Prof Williams says.

Science fiction writer Isaac Asimov formulated his own laws of robotics which stated that robots aren't allowed to harm humans.

Easier said than done, says Prof Williams.

She points out that "you need a lot of cognitive capability to determine harm if you're in a different kind of body."

"What will we do when we have to deal with entities . . . who have perceptions beyond our own and can reason as well as we can, or potentially better?" she says.

"It becomes very interesting.

"There are already robotic soldiers (on wheels) in Iraq trained to kill people."

But surely if we build these robots we can program them not to harm us?

Not so, says Prof Williams.

"One of the things we'll want robots to do is communicate," she says.

"But in order to have a conversation you need the capability to build a mental model of the person you're communicating with.

"And if you can model other people or other systems' cognitive abilities then you can deceive."

At the moment ASIMO has only limited AI.

He is not really aware of what he's doing—like the robots Prof Williams and her UTS team build to play soccer at international RoboCup competitions.

The goal of the annual competition is to develop a team of humanoids that can beat the official FIFA human world cup champions by 2050.

The next one will be held in China 2008.

Until this year, four-legged Sony AIBO's took to the field, but now games are played between teams of Aldebaran Nao humanoids.

Prof Williams says the robots are taught "behaviours" which allow them to make decisions by themselves.

"If a robot doesn't have the ball then he searches for the ball," she says.

"Then when he finds the ball that triggers a new behaviour which says 'go to the ball.'

"Then when he's at the ball he goes into another behaviour which is 'grab the ball,' and then the next behaviour is 'find the goal' and the next behaviour is 'kick.'"

The robots can learn from past experience, but Prof Williams says the real challenge is teaching them what to forget.

"If you remember everything it's very hard to then retrieve relevant information on the fly when it's needed."

ASIMO can interpret the postures and gestures of humans and move independently in response, as well as remember different voices and faces.

Dr Caroline West, a senior lecturer in philosophy at Sydney University, says we should already be thinking about what will happen when humanoids develop the ability to reason and integrate into society.

She says there are legitimate concerns about creating "terminator-type warrior robots," but there are broader ethical issues as well.

For example, if humanoids become as intelligent and capable of feeling as humans, should they be given the same rights?

The question cuts to the heart of what a "person" is.

"If something is a person then it has serious rights, and what it takes to be a person is to be self-conscious and able to reason," she says.

"If silicon-based creatures get to have those abilities then they would have the same moral standing as persons.

"Just as we think it's not okay to enslave persons so it would be wrong to enslave these robots if they really were self-conscious."

ASIMO operator Ekkasit Najaitrvek insists a humanoid is much more than a super computer.

"He's a person," Mr Najaitrvek says of ASIMO.

"Sometimes he's like a kid. Most of the time he's my friend.

"You can see when he looks at you—you can't deny he's a person."

Robot Babies*

By Abigail Tucker
Smithsonian, July 2009

Einstein the robot has enchanting eyes, the color of honey in sunlight. They are fringed with drugstore-variety false eyelashes and framed by matted gray brows made from real human hair. "What is that, makeup?" a visiting engineer asks, and, indeed, on closer examination I can see black eyeliner smeared beneath Einstein's lower lids, à la David Bowie in 1971. The machine's gaze is expressive—soulful, almost.

David Hanson, Einstein's creator, is visiting from Texas to help scientists here at the University of California at San Diego (UCSD) prepare the robot for an upcoming conference. Hanson switches the robot on—really just a head and neck—and runs it through some of its dozens of expressions. Its lips purse. Its brow furrows. Its eyes widen as though in horror, then scrunch mirthfully as it flashes a grin. The 27 motors in the face make a wretched grinding sound, and when the mouth opens, I see a tangle of wires where the prodigious brain should be. Einstein's white wig is missing and the skin of its neck hangs in flaps, because its shoulders, made of plastic, got shattered in shipping.

Still, the effect is so lifelike that even jaded graduate students have stopped by to stroke the robot's wrinkled cheek, which is encased in a substance called Frubber—"face rubber," Hanson explains—that feels buttery soft and cold as a cadaver. Engineers working at night have been so unnerved by Einstein's presence they asked that it be covered up.

That delights Javier Movellan, leader of the university group that recently spent $70,000 in research grants to buy the robotic head from Hanson. Einstein will eventually be able to track people's movements with cameras that twinkle behind its pupils and judge their gender, age (well, whether they're over or under 40), if they wear glasses or have facial hair and when they're smiling. The robot is also equipped with conversational software and can hold forth while staring deep into a person's eyes. "Do you know any jokes?" Hanson asks the machine.

Einstein: Why did the chicken cross the road?

Hanson: Tell me about artificial intelligence.

Einstein: I feel real, not artificial.

Hanson: I want to hear more about the future of robotics.

Einstein: Robots will become ever more like animals and people. Robots will continue to get more amazing and cool.

Einstein is the product of a remarkable collaboration. Hanson, a robot designer and the founder of the Dallas-based firm Hanson Robotics, has used classical sculpting techniques to animate robotic likenesses of Philip K. Dick, author of *Do Androids Dream of Electric Sheep?* (the basis of the apocalyptic movie *Blade Runner*), his own wife (he had to use a male skull model, "which masculinized her a bit") and more than a dozen other people. Movellan, a psychologist and software pioneer who runs UCSD's Machine Perception Laboratory, develops technology that approximates human senses. Einstein is, at present, a research tool to explore how a machine can perceive and react to human facial expressions; that capacity could later have many practical applications in entertainment and education, alerting the robot teachers of the future, say, that their human pupils are daydreaming.

For the most part, though, the intelligence I perceived in Einstein—its intense eye contact, its articulate soliloquies—was an illusion. Its answers to questions were canned and its interpretive powers were extremely limited. In short, Einstein is no Einstein. Overall, robots can do amazing things—play the violin, dismantle bombs, fire missiles, diagnose diseases, tend tomato plants, dance—but they sorely lack the basics. They recite jokes but don't get them. They can't summarize a movie. They can't tie their shoelaces. Because of such shortcomings, whenever we encounter them in the flesh, or Frubber, as it were, they are bound to disappoint.

Rodney Brooks, an M.I.T. computer scientist who masterminded a series of robotics innovations in the 1990s, said recently that for a robot to have truly human-like intelligence, it would need the object-recognition skills of a 2-year-old child, the language capabilities of a 4-year-old, the manual dexterity of a 6-year-old and the social understanding of an 8-year-old. Experts say they are far from reaching those goals. In fact, the problems that now confound robot programmers are puzzles that human infants often solve before their first birthday How to reach for an object. How to identify a few individuals. How to tell a stuffed animal from a bottle of formula. In babies, these skills are not preprogrammed, as were the perceptual and conversational tricks Einstein showed me, but rather are cultivated through interactions with people and the environment.

But what if a robot could develop that way? What if a machine could learn like a child, as it goes along? Armed with a nearly $3 million National Science Foundation grant, Movellan is now tackling that very question, leading a team of cognitive scientists, engineers, developmental psychologists and roboticists from UCSD and beyond. Their experiment—called Project One, because it focuses on the first year of development—is a wildly ambitious effort to crack the secrets of human intelligence. It involves, their grant proposal says, "an integrated system . . . whose sensors and actuators approximate the levels of complexity of human infants."

In other words, a baby robot.

The word "Robot" hit the world stage in 1921, in the Czech science fiction writer Karel Capek's play *Rossum's Universal Robots*, about a factory that creates artificial people. The root is the Czech robota, for serf labor or drudgery. Broadly understood, a robot is a machine that can be programmed to interact with its surroundings, usually to do physical work.

We may associate robots with artificial intelligence, which uses powerful computers to solve big problems, but robots are not usually designed with such lofty aspirations; we might dream of Rosie, the chatty robot housekeeper on "The Jetsons," but for now we're stuck with Roomba, the disk-shaped, commercially available autonomous vacuum cleaner. The first industrial robot, called Unimate, was installed in a General Motors factory in 1961 to stack hot pieces of metal from a die-casting machine. Today, most of the world's estimated 6.5 million robots perform similarly mundane industrial jobs or domestic chores, though 2 million plug away at more whimsical tasks, like mixing cocktails. "Does [the robot] prepare the drink with style or dramatic flair?" ask the judging guide-lines for the annual RoboGames bar-tending competition, held in San Francisco this summer. "Can it prepare more than a martini?"

Now imagine a bartender robot that could waggle its eyebrows sympathetically as you pour out the story of your messy divorce. Increasingly, the labor we want from robots involves social fluency, conversational skill and a convincing humanlike presence. Such machines, known as social robots, are on the horizon in health care, law enforcement, child care and entertainment, where they might work in concert with other robots and human supervisors. Someday, they might assist the blind; they've already coached dieters in an experiment in Boston. The South Korean government has said it aims to have a robot working in every home by 2020.

Part of the new emphasis on social functioning reflects the changing economies of the richest nations, where manufacturing has declined and service industries are increasingly important. Not coincidentally, societies with low birthrates and long life expectancies, notably Japan, are pushing hardest for social robots, which may be called upon to stand in for young people and perform a wide variety of jobs, including caring for and comforting the old.

Some scientists working on social robots, like Movellan and his team, borrow readily from developmental psychology A machine might acquire skills as a human child does by starting with a few basic tasks and gradually constructing a more sophisticated competence—"bootstrapping," in scientific parlance. In contrast to preprogramming a robot to perform a fixed set of actions, endowing a robot computer with the capacity to acquire skills gradually in response to the environment might produce smarter, more human robots.

"If you want to build an intelligent system, you have to build a system that becomes intelligent," says Giulio Sandini, a bioengineer specializing in social robots at the Italian Institute of Technology in Genoa. "Intelligence is not only what you know but how you learn more from what you know. Intelligence is acquiring information, a dynamic process."

"This is the brains!" Movellan shouted over the din of cyclone-strength air conditioners. He was pointing at a stack of computers about ten feet tall and six feet deep, sporting dozens of blinking blue lights and a single ominous orange one. Because the Project One robot's metal cranium will not be able to hold all the information-processing hardware that it will need, the robot will be connected by fiber-optic cables to these computers in the basement of a building on the UCSD campus in La Jolla. The room, filled with towering computers that would overheat if the space weren't kept as cold as a meat locker, looks like something out of *2001: A Space Odyssey.*

As Einstein could tell you, Movellan is over 40, bespectacled and beardless. But Einstein has no way of knowing that Movellan has bright eyes and a bulky chin, is the adoring father of an 11-year-old daughter and an 8-year-old son and speaks English with an accent reflecting his Spanish origins.

Movellan grew up amid the wheat fields of Palencia, Spain, the son of an apple farmer. Surrounded by animals, he spent endless hours wondering how their minds worked. "I asked my mother, 'Do dogs think? Do rats think?'" he says. "I was fascinated by things that think but have no language."

He also acquired a farm boy's knack for working with his hands; he recalls that his grandmother scolded him for dissecting her kitchen appliances. Enamored of the nameless robot from the 1960s television show "Lost in Space," he built his first humanoid when he was about 10, using "food cans, light bulbs and a tape recorder," he says. The robot, which had a money slot, would demand the equivalent of $100. As Movellan anticipated, people usually forked over much less. "That's not $100!" the robot's prerecorded voice would bellow. Ever the mischievous tinkerer, he drew fire 30 years later from his La Jolla homeowners association for welding robots in his garage.

He got his PhD in developmental psychology at the University of California at Berkeley in 1989 and moved on to Carnegie Mellon University, in Pittsburgh, to conduct artificial intelligence research. "The people I knew were not really working on social robots," he says. "They were working on vehicles to go to Mars. It didn't really appeal to me. I always felt robotics and psychology should be more together than they originally were." It was after he went to UCSD in 1992 that he began working on replicating human senses in machines.

A turning point came in 2002, when he was living with his family in Kyoto, Japan, and working in a government robotics lab to program a long-armed social robot named Robovie. He hadn't yet had much exposure to the latest social robots and initially found them somewhat annoying. "They would say things like, 'I'm lonely, please hug me,'" Movellan recalls. But the Japanese scientists warned him that Robovie was special. "They would say 'you'll feel something.' Well, I dismissed it—until I felt something. The robot kept talking to me. The robot looked up at me and, for a moment, I swear this robot was alive."

Then Robovie enfolded him in a hug and suddenly—"magic," says Movellan.

"This is something I was unprepared for from a scientific point of view. This intense feeling caught me off guard. I thought, Why is my brain put together so that this machine got me? Magic is when the robot is looking at things and you reflexively want to look in the same direction as the robot. When the robot is looking at you instead of through you. It's a feeling that comes and goes. We don't know how to make it happen. But we have all the ingredients to make it happen."

Eager to understand this curious reaction, Movellan introduced Robovie to his 2-year-old son's preschool class. But there the robot cast a different spell. "It was a big disaster," Movellan remembers, shaking his head. "It was horrible. It was one of the worst days of my life." The toddlers were terrified of Robovie, who was about the size of a 12-year-old. They ran away from it screaming.

That night, his son had a nightmare. Movellan heard him muttering Japanese in his sleep: "Kowai, kowai." Scary, scary.

Back in California, Movellan assembled, in consultation with his son, a kid-friendly robot named RUBI that was more appropriate for visits to toddler classrooms. It was an early version of the smiling little machine that stands sentinel in the laboratory today, wearing a jaunty orange Harley-Davidson bandanna and New Balance sneakers, its head swiveling in an inquisitive manner. It has coasters for eyes and a metal briefcase for a body that snaps open to reveal a bellyful of motors and wires.

"We have learned a lot from this little baby," Movellan said, giving the robot an affectionate pat on its square cheek.

For the past several years he has embedded RUBI at a university preschool to study how the toddlers respond. Various versions of RUBI (some of them autonomous and others puppeteered by humans) have performed different tasks. One taught vocabulary words. Another accompanied the class on nature walks. (That model was not a success; with its big wheels and powerful motors, RUBI swelled to an intimidating 300 pounds. The kids were wary, and Movellan was, too.)

The project has had its triumphs—the kids improved their vocabularies playing word games displayed on RUBI's stomach screen—but there have been setbacks. The children destroyed a fancy robotic arm that had taken Movellan and his students three months to build, and RUBI's face detector consistently confused Thomas the Tank Engine with a person. Programming in incremental fixes for these problems proved frustrating for the scientists. "To survive in a social environment, to sustain interaction with people, you can't possibly have everything preprogrammed," Movellan says.

Those magic moments when a machine seems to share in our reality can sometimes be achieved by brute computing force. For instance, Einstein's smile-detection system, a version of which is also used in some cameras, was shown tens of thousands of photographs of faces that had been marked "smiling" or "not smiling." After cataloging those images and discerning a pattern, Einstein's computer can "see" whether you are smiling, and to what degree. When its voice software is cued to compliment your pretty smile or ask why you look sad, you might feel a spark of unexpected emotion.

But this laborious analysis of spoon-fed data—called "supervised learning"—is nothing like the way human babies actually learn. "When you're little nobody points out ten thousand faces and says 'This is happy, this is not happy, this is the left eye, this is the right eye,'" said Nicholas Butko, a PhD student in Movellan's group. (As an undergraduate, he was sentenced to labeling a seemingly infinite number of photographs for a computer face-recognition system.) Yet babies are somehow able to glean what a human face is, what a smile signifies and that a certain pattern of light and shadow is Mommy.

To show me how the Project One robot might learn like an infant, Butko introduced me to Bev, actually BEV, as in Baby's Eye View. I had seen Bev slumped on a shelf above Butko's desk without realizing that the Toys 'R' Us-bought baby doll was a primitive robot. Then I noticed the camera planted in the middle of Bev's forehead, like a third eye, and the microphone and speaker under its purple T-shirt, which read, "Have Fun."

In one experiment, the robot was programmed to monitor noise in a room that people periodically entered. They'd been taught to interact with the robot, which was tethered to a laptop. Every now and then, Bev emitted a babylike cry. Whenever someone made a sound in response, the robot's camera snapped a picture. The robot sometimes took a picture if it heard no sound in response to its cry, whether or not there was a person in the room. The robot processed those images and quickly discerned that some pictures—usually those taken when it heard a response—included objects (faces and bodies) not present in other pictures. Although the robot had previously been given no information about human beings (not even that such things existed), it learned within six minutes how to tell when someone was in the room. In a remarkably short time, Bev had "discovered" people.

A similar process of "unsupervised learning" is at the heart of Project One. But Project One's robot will be much more physically sophisticated than Bev—it will be able to move its limbs, train its cameras on "interesting" stimuli and receive readings from sensors throughout its body—which will enable it to borrow more behavior strategies from real infants, such as how to communicate with a caregiver. For example, Project One researchers plan to study human babies playing peekaboo and other games with their mothers in a lab. Millisecond by millisecond, the researchers will analyze the babies' movements and reactions. This data will be used to develop theories and eventually programs to engineer similar behaviors in the robot.

It's even harder than it sounds; playing peekaboo requires a relatively nuanced understanding of "others." "We know it's a hell of a problem," says Movellan. "This is the kind of intelligence we're absolutely baffled by. What's amazing is that infants effortlessly solve it." In children, such learning is mediated by the countless connections that brain cells, or neurons, form with one another. In the Project One robot and others, the software itself is formulated to mimic "neural networks" like those in the brain, and the theory is that the robot will be able to learn new things virtually on its own.

The robot baby will be able to touch, grab and shake objects, and the researchers hope that it will be able to "discover" as many as 100 different objects that infants might encounter, from toys to caregivers' hands, and figure out how to manipulate them. The subtleties are numerous; it will need to figure out that, say, a red rattle and a red bottle are different things and that a red rattle and a blue rattle are essentially the same. The researchers also want the robot to learn to crawl and ultimately walk.

Perhaps the team's grandest goal is to give the robot the capacity to signal for a caregiver to retrieve an object beyond its grasp. Movellan calls this the "Vygotsky reach," after developmental psychologist Lev Vygotsky who identified the movement—which typically occurs when a child is about a year old—as an intellectual breakthrough, a transition from simple sensory-motor intelligence to symbolic intelligence. If the scientists are successful, it will be the first spontaneous symbolic gesture by a robot. It will also be a curious role reversal—the robot commanding the human, instead of vice versa.

"That's a pretty important transition," says Jonathan Plucker, a cognitive scientist at Indiana University who studies human intelligence and creativity. Plucker had no prior knowledge of Project One and its goals, but he was fresh from watching the season finale of "Battlestar Galactica," which had left him leery of the quest to build intelligent robots. "My sense is that it wouldn't be hard to have a robot that reaches for certain types of objects," he says, "but it's a big leap to have a machine that realizes it wants to reach for something and uses another object, a caregiver, as a tool. That is a much, much more complex psychological process."

At present, the Project One robot is all brains. While the big computer hums in its air-conditioned cavern, the body is being designed and assembled in a factory in Japan.

Construction is expected to take about nine months.

A prototype of the PROJECT ONE robot body already exists, in the Osaka laboratory of Hiroshi Ishiguro, the legendary Japanese roboticist who, in addition to creating Robovie, fashioned a robotic double of himself, named Geminoid, as well as a mechanical twin of his 4-year-old daughter, which he calls "my daughter's copy" ("My daughter didn't like my daughter's copy" he told me over the phone. "Its movement was very like a zombie." Upon seeing it, his daughter—the original—cried.) Ishiguro's baby robot is called the Child-Robot with Biomimetic Body, or CB2 for short. If you search for "creepy robot baby" on YouTube, you can see clips of four-foot-tall CB2 in action. Its silicone skin has a grayish cast; its blank, black eyes dart back and forth. When first unveiled in 2007, it could do little more than writhe, albeit in a very babylike way and make pathetic vowel sounds out of the tube of silicone that is its throat.

"It has this ghostly gaze," says Ian Fasel, a University of Arizona computer scientist and a former student of Movellan's who has worked on the Japanese project. "My friends who see it tell me to please put it out of its misery. It was often lying on the floor of the lab, flopping around. It gives you this feeling that it's struggling to be a real boy but it doesn't know how."

When Movellan first saw CB2, last fall, as he was shopping around for a Project One body, he was dismayed by the lack of progress the Japanese scientists had made in getting it to move in a purposeful way. "My first impression was that there was no way we would choose that robot," Movellan recalls. "Maybe this robot is impossible to control. If you were God himself, could you control it?"

Still, he couldn't deny that the CB2 was an exquisite piece of engineering. There have been other explicitly childlike robots over the years—creations such as Babybot and Infanoid—but none approach CB2's level of realism. Its skin is packed with sensors to collect data. Its metal skeleton and piston-driven muscles are limber, like a person's, not stiff like most robots', and highly interconnected: if an arm moves, motors in the torso and elsewhere respond. In the end, Movellan chose CB2.

The body's human-ness would help the scientists develop more brainlike software, Movellan decided. "We could have chosen a robot that could already do a lot of the things we want it to do—use a standard robotic arm, for instance," Movellan says. "Yet we felt it was a good experiment in learning to control a more biologically inspired body that approximates how muscles work. Starting with an arm more like a real arm is going to teach us more."

The Project One team has requested tweaks in CB2's design, to build in more powerful muscles that Movellan hopes will give it the strength to walk on its own, which the Japanese scientists—who are busy developing a new model of their own—now realize the first CB2 will never do. Movellan is also doing away with the skin suit, which sometimes provides muddled readings, opting instead for a Terminator-like metal skeleton encased in clear plastic. ("You can always put clothes on," Movellan reasons.) He had hoped to make the robot small enough to cradle, but the Japanese designers told him that is currently impossible. The baby will arrive standing about three feet tall and weighing 150 pounds.

What a social robot's face should look like is a critical, and surprisingly difficult, decision. CB2's face is intended to be androgynous and abstract, but somehow it has tumbled into what robotics experts term the "uncanny valley," where a machine looks just human enough to be unsettling. The iCub, another precocious child-inspired robot being built by a pan-European team, looks more appealing, with cartoonish wide eyes and an endearing expression. "We told the designers to make it look like someone who needed help," says the Italian Institute of Technology's Sandini, who's leading the project. "Someone . . . a little sad."

When I met Movellan he seemed flummoxed by the matter of his robot's facial appearance: Should the features be skeletal or soft-tissue, like Einstein's? He was also pondering whether it would be male or female. "All my robots so far have been girls—my daughter has insisted," he explains. "Maybe it's time for a boy." Later, he and his coworkers asked Hanson to help design a face for the Project One robot, which will be named Diego. The "developmental android" will be modeled after a real child, the chubbycheeked nephew of a researcher in Movellan's lab.

Though Movellan believes that a human infant is born with very little preexisting knowledge, even he says it comes with needs: to be fed, warmed, napped and relieved of a dirty diaper. Those would have to be programmed into the robot, which quickly gets complicated. "Will this robot need to evacuate?" says John Watson, a University of California at Berkeley professor emeritus of psychology who is a Project One consultant. "Will the thing need sleep cycles? We don't know."

Others outside the project are skeptical that baby robots will reveal much about human learning, if only because a human grows physically as well as cognitively. "To mimic infant development, robots are going to have to change their morphology in ways that the technology isn't up to," says Ron Chrisley, a cognitive scientist at the University of Sussex in England. He says realistic human features are usually little more than clever distractions: scientists should focus on more basic models that teach us about the nature of intelligence. Human beings learned to fly, Chrisley notes, when we mastered aerodynamics, not when we fashioned realistic-looking birds. A socially capable robot might not resemble a human being anymore than an airplane looks like a sparrow.

Maybe the real magic of big-eyed, round-faced robobabies is their ability to manipulate our own brains, says Hamid Ekbia, a cognitive science professor at Indiana University and the author of *Artificial Dreams: The Quest for Non-Biological Intelligence*. Infantalized facial features, he says, primarily tap into our attraction to cute kids. "These robots say more about us than they do about machines," says Ekbia. "When people interact with these robots, they get fascinated, but they read beneath the surface. They attribute qualities to the robot that it doesn't have. This is our disposition as human beings: to read more than there is."

Of course, Movellan would counter that such fascination is, in Project One's case, quite essential: to develop like a real child, the machine must be treated like one.

Each Project One researcher defines success differently Some will declare victory if the robot learns to crawl or to identify basic objects. Watson says he would be grateful to simulate the first three months of development. Certainly, no one expects the robot to progress at the same rate as a child. Project One's timeline extends over four years, and it may take that long before the robot is exposed to people outside the lab—"caregivers" (read: undergrads) who will be paid to baby-sit. Lacking a nursery, the robot will be kept behind glass on a floor beneath Movellan's lab, accessible, for the time being, only to researchers.

As for Movellan, he hopes that the project will "change the way we see human development and bring a more computational bent to it, so we appreciate the problems the infant brain is solving." A more defined understanding of babies' brains might also give rise to new approaches to developmental disorders. "To change the questions that psychologists are asking—that to me is the dream," Movellan adds. "For now it is, how do you get its arm to work, the leg to work? But when we put the pieces together, things will really start to happen."

Before leaving the lab, I stop to bid goodbye to Einstein. All is not well with the robot. Its eye cameras have become obsessed with the glowing red exit sign over the workshop's door. Hanson switches the robot off and on; its movements are palsied; its eyes roll. Its German accent isn't working and the tinny-sounding conversational software seems to be on the fritz. Hanson peers into its eyes. "Hi there," he says. "Can you hear me? Are you listening?"

Einstein: (No response.)

Hanson: Let's get into the topic of compassion.

Einstein: I don't have good peripheral vision.

Einstein: (Continuing.) I am just a child. I have a lot to learn, like what it is to truly love.

Students working nearby are singing along to a radio blasting Tina Turner's "What's Love Got to Do With It," oblivious to Einstein's plight. For me, though, there is something almost uncomfortable about watching the robot malfunction, like seeing a stranger struggle with heavy suitcases. Does this count as magic?

On a worktable nearby, something catches my eye. It is a copy of a Renaissance-era portrait of Mary and the infant Jesus—Carlo Crivelli's *Madonna con Bambino*, the engineers say, which another robot in the room is using to practice analyzing images. The painting is the last thing I expect to see among the piles of tools and snarls of wires, but it occurs to me that building a humanoid robot is also a kind of virgin birth. The child in the painting is tiny but already standing on its own. Mary's eyes are downcast and appear troubled; the baby stretches one foot forward, as though to walk, and gazes up.

Robots May Soon Have Tails, Whiskers*

Chinadaily.com, June 10, 2009

Agnes Guillot dreams of one day seeing a giant white rat called Psikharpax scuttling fearlessly around her lab. When she does, it will be time to scream—but from joy rather than fear for it could be a turning point in the history of robotics.

Psikharpax—named after a cunning king of the rats, according to a tale attributed to Homer—is the brainchild of European researchers who believe it may push back a frontier in artificial intelligence.

Scientists have strived for decades to make a robot that can do more than make repetitive, programmed gestures. These are fine for manufacturing cars or amusing small children, but are of little help in the real world.

One of the biggest obstacles is learning ability. Without the smarts to figure out dangers and opportunities, a robot is helpless without human intervention.

Rather than try to replicate human intelligence it would be better to start at the bottom and figure out simpler abilities that humans share with other animals, they say. These include navigating, seeking food and avoiding dangers.

Rat robots are being built in other labs in Britain, the US and elsewhere. Two years ago, for instance, a team at the ITAM technical institute in Mexico City reprogrammed a Sony Aibo dog using rat-simulated sofware.

But the European researchers believe that Psikharpax is unique in its biomimickry, sophistication of sensors and controls and software based on rat neurology.

Their artificial rodent has two cameras for eyes, two microphones for ears and tiny wheels, driven by a battery-powered motor, to provide movement. A couple of dozen whiskers stretch out impressively either side of its long, pointed snout.

Data from these artificial organs goes to Psikharpax's "brain," a chip whose software hierarchy mimicks the structures in a rat's brain that process and analyze what is seen, heard and sensed.

The goal is to get Psikharpax to be able to "survive" in new environments. It would be able to detect and move around things in its way and spot an opportunity for "feeding"—recharging its battery at power points placed around the lab.

"We want to make robots that are able to look after themselves and depend on humans as little as possible," said Guillot.

Dreaming of Electric Sheep*

By Steven Johnson
The Nation, September 3–10, 2001

Ever doubt that you are living in the twenty-first century? Here's a sure-fire cure: Pay a visit to the website for the Society of Robotic Combat, where you'll encounter the following, utterly earnest, mission statement: "The Society of Robotic Combat seeks to serve the competition robot community in two ways. First, we will represent the builders by creating and maintaining an equitable rule set for robot competitions. Next we will strive to assemble and disseminate the premier knowledge base for the practical construction and operation of competitive robots."

You may well be surprised to learn that there is such a thing as "robot competition"—and perhaps even more surprised to find that these competitions have advanced far enough to require a governing body and "equitable rule sets." Indeed, you can see those rules in action weeknights on Comedy Central, which has just begun showing new episodes of its cult hit *BattleBots*—a kind of QVC-meets-*Gladiator* hybrid where the viewer is introduced to a parade of elaborate mechanical devices with names like Killerhurtz and Spike, which then proceed to bludgeon one another into oblivion in a Plexiglas-encased coliseum, while screaming fans cheer them on.

BattleBots and the Society of Robotic Combat might look at first glance like they just dropped out of a Philip K. Dick novel, particularly if your previous notion of mechanized sporting was Casey Martin's golf cart. But the competitive robots of *BattleBots*—or Europe's insanely popular series *Robot Wars*—aren't quite as advanced as you might initially think. The first time I tuned into *BattleBots*, I sat entranced for ten minutes as the two machines engaged in what seemed like a remarkably sophisticated dance for a pair of glorified lawnmowers. I thought to myself, When did the robots get this smart? And then the screen cut to the sidelines, where their human owners furiously manipulated their remote-control devices.

* "Dreaming of Electric Sheep" by Steven Johnson. Reprinted with permission from the September 3-10, 2001 issue of *The Nation*. For subscription information, call 1-800-333-8536. Portions of each week's *Nation* magazine can be accessed at http://www.thenation.com.

Killerhurtz and Spike might have been robotic gladiators, but they were worse than slaves of their human masters—they were closer to high-tech puppets.

But hey, it's only 2001. By 2005, we'll be watching autonomous, semi-intelligent machines dueling one another without the remote-control devices, and before long one of those machines will spontaneously generate its own rope-a-dope strategy, or perhaps even thrust a Black Power fist into the sky, and it will become our first real cyborg hero. How long, then, until that machine's achievements start to be recorded by the sports pages, alongside the nonhuman athletes of the Belmont Stakes and the Indy 500? And if that machine should pass convincingly into the realm of human sport, how long until its growing intelligence starts to become a threat to its creators?

That is the question that hangs over the middle section of *AI*, Steven Spielberg's erratic and commercially unsuccessful collaboration with the late Stanley Kubrick. Set several hundred years in the future, in a world populated by intelligent, adaptive mechanical life forms called Mecha, the film begins with the creation of a Mecha child—played by the sublime Haley Joel Osment—capable of feeling love for an adopted human mother. A young couple, whose first son lies trapped in a comalike state of suspended animation because of some unnamed illness, takes the roboboy David home, and slowly a bond develops between mother and adopted son—until the real son wakes from his coma and returns home to torture his new kid brother. Before long, David has been exiled from his human family and left to wander the earth like a cyborg Ronin.

"I'm sorry I didn't teach you more about the world," his mother sobs before abandoning him in the obligatory fog-shrouded pine forest. She's got a good reason to be sorry—the world turns out to harbor a growing subculture of anti-Mecha fury, localized around traveling Flesh Fairs, during which "unlicensed"—but still functioning—old robots are dismembered, disemboweled and burned with a spectacular violence that brings to mind the opening pages of *Discipline and Punish*. If *BattleBots* proposes the question of what happens when machines get smart enough to entertain us, the Flesh Fairs of *AI* suggest what will happen when they get too smart. The sequence is as dark and disturbing as anything Spielberg has shot, and its politics are surprisingly complex. As the Mecha are ritually destroyed, the PA system blasts slogans about ending corporate artificiality while a death metal band thunders in the background—blur your eyes a little and you might be at a Rage Against the Machine show. When David is thrust upon the stage, pleading for his life in a 10-year-old's terrified voice, the crowd abruptly turns on the Flesh Fair's proprietors. The message seems to be that it's inhuman to torture a nonhuman who simulates human emotion convincingly enough. David fakes so real he is beyond fake.

But if *BattleBots* and the Unabomber manifesto make the anti-Mecha carnage seem strikingly close to home, in another sense, David's ontological crisis is an old story. We've been obsessed with the emotional lives of robots for almost as long as we've been obsessed with robots themselves. Think of *Blade Runner*'s original title, *Do Androids Dream of Electric Sheep?*, or Kubrick's paranoid HAL 9000. ("I enjoy working with people," HAL tells an interviewer halfway through *2001*, when asked if he ever gets frustrated. "I have a stimulating relationship with Dr. Poole and Dr. Bowman.") The anxiety about machine intelligence has always centered on the fear that the machines will resemble us too closely, that the lines between Mecha and non-Mecha will blur. (Hence the longstanding debate over whether Harrison Ford's character in *Blade Runner* is a replicant.) Spielberg and Kubrick's twist is to take that subterfuge to the next level: What would happen if a machine could mimic—or perhaps even experience—the fundamental human emotion of maternal love?

The connection between emotion and artificial intelligence may be an old literary trope, but it is very much au courant in the scientific and technology worlds. A number of recent books by leading neuroscientists have argued that emotion plays an integral role in creating consciousness, including Susan Greenfield's *The Private Life of the Brain* and Antonio Damasio's *The Feeling of What Happens*. In a 1999 bestseller, technovisionary Ray Kurzweil predicts the rise of "spiritual" machines sometime this century. Greenfield paraphrases MIT's legendary artificial intelligence researcher Marvin Minsky: "The important question is not to ask whether an intelligent machine could have emotions, but whether a machine could be truly intelligent without them." All the leading indicators suggest that the next great world-transforming technology revolution will be genomic, but the wave due to roll in after that one looks to be the rise of genuine artificial intelligence. Can the Flesh Fairs be far behind?

If you believe the self-conscious fairy tales of *AI*, that second wave will do as much damage as the rising tides of global warming, which in Spielberg's movie have drowned Manhattan and left the industrialized world coping with population controls by inventing robot children to satisfy unmet parental urges. According to the Spielberg/Kubrick account, emotionally nuanced AI will turn the world upside down, for precisely the reasons that the robots have always unnerved us in the folklore: They'll act too much like the way we do. But the reality is likely to be more uncanny than that, and harder to predict. It does seem probable that machines will learn to interact with us in genuinely adaptive and improvisational ways sometime over the next hundred years; will learn to recognize our subtle failings and strengths—even learn how to learn on their own. But the disturbing thing about that revolution is not likely to be smart machines behaving too much like humans. Instead, it will almost certainly be their strangeness. This is where Spielberg missed a great opportunity to reinvent the thinking-machine genre, and perhaps that missed opportunity partly accounts for the frustration so many viewers felt sitting through *AI*.

There's a now-familiar riff about extraterrestrial life that maintains that the "little green men" of 1950s lore—or the cone-headed oversized infants of recent fare—are the ultimate in Homo sapiens provincialism: When intelligent life arrives from outer space, it won't look like bipedal primates—it'll look like a cloud or a cluster of bacteria or something so different from our earthbound life forms that we won't even perceive it. (Marlon Brando was on a parallel wavelength when he famously proposed playing Superman's father as a green suitcase.) We should expect the same shock of unfamiliarity when confronting the first generation of true artificial intelligence. The first thinking machines will be smart, but they won't think the way we do; and if they experience emotion, it will be dramatically unlike any emotion we've experienced in our human consciousness.

This gap between machine and human intelligence will arise because the first generation of true AI will almost certainly be the product of evolution, not traditional engineering. A structure as complex as the human brain—which may well be the most complex biological system on the planet—can't simply be replicated *in silico*; you can't draw up a blueprint for the myriad interconnections of our billions of neurons. But in a matter of decades, as Ray Kurzweil has convincingly shown, we will have machines capable of doing as many parallel calculations per second as the human brain, and by the end of the century, our brains will look like Palm Pilots next to our most advanced supercomputers. But it's unlikely that we'll be able to sit down and write the software that would produce a thinking machine, because such a program would be mind-numbingly complicated to design. We'll have kitchen supplies for cooking up artificial intelligence, in other words—we just won't have a recipe.

And so the AI pioneers of the twenty-first century will most likely rely on a kind of artificial natural selection to create their thinking machines. A bank of massively parallel supercomputers designed to simulate the multiple connections of our neuronal system will be connected to some kind of external world, and programmed to experiment with more or less random responses to the data coming in from that world. The humans will establish criteria that reward behavior that shows signs of intelligence and punish less promising behavior. Innovative new strategies generated by a computer will be preserved and shared with all the other machines; less successful strategies will be eliminated. Over time, the virtual gene pool will start to accumulate intelligent strategies for dealing with—and predicting events in—the outside world. A machine connected to some kind of light detector might begin to register the twenty-four-hour cycle of light and dark; that "insight" would then be passed on to the machine's virtual descendants, one of whom might then develop an awareness of seasonal cycles, by noting changes in the length of days. That strategy would be passed down to future generations, where it could be enhanced or expanded by new innovations. Presumably, somewhere at the end of this assisted evolution, a digital Copernicus awaits us.

For every successful assessment of the world, of course, there are a million failures, just as there are in the real-world fossil record. The difference here is that the pace of evolutionary time has been sped up dramatically; the supercomputers can churn through thousands of generations of artificial minds in a matter of minutes, if required. Already, this type of evolutionary approach has "bred" software programs that are faster and more stable than programs written using conventional techniques. For a single, focused task like number sorting or fingerprint recognition, these evolutionary techniques can be extremely effective. But for a more open-ended, multidimensional goal like general intelligence, the problem with the evolutionary approach—if problem is the right word for it—is that the human overseers don't have much control over the direction in which the software will evolve. You can push it in certain directions, particularly in the early stages of evolution, but before long it is bound to develop a mind of its own. Already, natural selection has seen fit to evolve countless forms of intelligence on the planet. If evolution now turns its invisible hand to growing machine intelligence, it seems unlikely that something resembling human smarts will emerge, no matter how much we rig the system.

So perhaps the sci-fi narratives have it the wrong way around: The sentient machines won't act like little boys or obedient butlers after all, but they'll be even more unnerving for their radical difference. (How multiculturalism will deal with the "new flesh" is a fascinating question.) Our creations won't be disappointed children, à la *Frankenstein* and *AI*, but rather some other species altogether—as familiar as a dolphin's mind, or whatever kind of distributed intelligence helps a termite colony build a nest. But if natural selection is to play such a crucial role in the creation of genuine AI, it's fair to ask about the environmental pressures that will shape the development of these new creatures. In what virtual petri dish will they do their learning? Where do you take a supercomputer when you want to teach it more about the world? One place to start might be that epic collection of human wisdom, hardcore porn and shameless self-promotion—the World Wide Web. A machine trying to bootstrap into intelligence might well begin by surveying those endless connected pages and soaking it all in. The web itself might not become self-aware, but perhaps a machine will cross that threshold by immersing itself in the datasphere for enough evolutionary cycles to start making sense of it all.

Although that machine intelligence would emerge out of studying our man-made information networks, there would be differences of kind, not just magnitude. Perhaps most important, it would be a thoroughly textual intelligence, having evolved in a universe of words. While language is certainly a central—and arguably defining—characteristic of human intelligence, it is a relatively new innovation. Our minds are also shaped by the ancient emotional centers of the limbic system and the advanced visual processing that we share with our primate cousins. (It's no accident that words for knowledge overlap so frequently with words for sight.) A machine trained to think for itself by reading data from web servers would lack the ballast and value judgments of our emotions, and the spatial logic that we borrow effortlessly from the shape recognizers of the visual system; the machine's smarts would be largely lexical, the intelligence you get when you live in a world of words (though perhaps a rudimentary visual intelligence could be evolved by studying the web's endless supply of Britney Spears pix). You can glimpse this lexical intelligence already in software applications like Kurzweil's Cybernetic Poet—a program that uses a kind of neural net technology called Markov models to detect textual patterns in poetry. Based on the patterns it associates with each poet, the program can compose its own poetry, in the style of John Ashbery or Robert Frost. (It can even compose authorial "blends.") The program seems to do best with elliptical and associative forms like haiku—it's not likely to churn out a cybernetic *Odyssey* anytime soon—but it suggests that word immersion can lead to remarkable advances. Yet if that immersion someday includes all the data stored on the web, not just the work of a few handpicked poets, the eventual results will vary wildly from the intelligence of the human race.

There's a striking premonition of this idea near the end of *2001*, as Dave Bowman shuts off the homicidal HAL 9000. As HAL winds down to nothingness—with his haunted, I'm-afraid-Dave refrain growing slower with each removed chip set—we learn that the state-of-the-art AI unit was "born" in 1992 in Urbana, Illinois. As it turns out, the modern web as we know it was born in Urbana in 1992, with the creation of the Mosaic browser. It's a fitting correlation: True artificial intelligence of the sort that HAL was supposed to embody may well evolve out of the web's vast archives. But that intelligence will not harbor the human failings—or strengths—that have historically been projected onto our fictional robot offspring. They will not love their mothers, nor dream of electric sheep.

Forecasts for Artificial Intelligence*

By Bohumir Stedron
The Futurist, March–April 2004

When the great Czech writer Karel Capek wrote his famous play *R.U.R. (Rossum's Universal Robots)* in 1920, he invented the word robot and gave the world a vision of machine intelligence. Within about 40 years, at the 1956 Dartmouth University Summer Research Project on Artificial Intelligence, the new scientific discipline Artificial Intelligence was established on the principle that human learning and intelligence could be precisely described and potentially simulated by machines.

AI development has now passed through three phases:

- Romantic period (1956–1965), whose best-known product, the General Problem Solver, performed various kinds of mathematical tasks.
- Ice age (1965–1980), when AI was not yet fully respected as a scientific discipline, but great progress was nevertheless made. Programming languages such as LISP and PROLOG were developed, and mathematical and logical models such as the Robinson resolution were discovered.
- Applications period (1980–present), when uses of AI techniques in military, industry, medicine, and other services have shown the financial sector that a new segment for investment is open.

What's next for artificial-intelligence development and applications? Here are a few possible scenarios incorporating new trends and discontinuities.

THE AGE OF MERGING (2010–2020)

- The United States, Germany, Japan, France, Ireland, Finland, and China represent the leading countries in high technology R&D.
- All laws enacted by legislatures in Washington and Brussels rely heavily on AI-based expert systems.

- Intelligent computers and telecommunication networks allow voice command for 3-D Internet, radio and television, mobile phones, medical care, and other services.
- Intelligent computers and telecommunication networks dominate the pedagogical process.
- As information technologies, biotechnologies, and nanotechnologies merge, so do the scientific disciplines developing them.
- Direct human-Internet communication is made possible with an implanted chip (later, without chips).
- New discoveries lead to quantum and DNA computers, as well as new materials incorporating low levels of intelligence.
- Antiviral programs destroy the occurrence of artificial life in order to avoid chaos like the blackouts that rocked Canada and the United States in August 2003.
- New laws are enacted to guarantee better health and vastly improved social well-being as AI development accelerates. For example, laws will protect against electromagnetic smog, regulate the use of home robots, protect data more stringently, prohibit the use of computer technologies in some segments of culture and arts (to avoid the domination of synthetic TV celebrities, for instance), and prohibit the construction of computer programs with self-preservation instincts.

THE AGE OF AI SELF-RELIANCE (2020–2030)

- Intelligent computers and telecommunications networks manage their own repairs, scientific research, and production.
- New materials incorporate high levels of intelligence.
- Direct communication among humans, computers, and cetaceans is possible using implanted chips (later, without chips).
- New laws recognize human rights for some robots.
- Intelligent networks are used to propel the moons Phobos and Deimos into the Martian polar caps, making the atmosphere more appropriate for humans.

THE NON-MYSTERIOUS AGE (2030–2040)

- A new, holographic model of the world replaces the geometrical one.
- AI systems mine holographic information from the environment.
- New explanations and applications emerge for mysterious phenomena, such as extrasensory perception and the use of energy fields in medicine and the military (bioenergy and/or psychotronics).

- AI makes it possible to create a copy of any human being's intellect; laws regulating and protecting these copies quickly follow.

How likely are these forecasts to bear fruit? Time will tell. The key issues to watch in AI development are not just the evolution of human beings and robotics, but also what is going on in the biosphere as AI enhances (or threatens) the futures of land animals, marine creatures, insects, and plants. And watch for the growing importance of artificial life—that is, the technological ability to mimic living behavior in systems such as the law, telecommunication networks, energy systems, and food production.

Bibliography

Books

Asimov, Isaac. *I, Robot.* New York: Doubleday, 1950.

Axe, David and Steve Olexa. *War Bots: How U.S. Military Robots Are Transforming War in Iraq, Afghanistan, and the Future.* Ann Arbor, Mich.: Nimble, 2008.

Bar-Cohen, Yoseph and David Hanson. *The Coming Robot Revolution: Expectations and Fears About Emerging Intelligent, Humanlike Machines.* New York: Springer, 2009.

Belfiore, Michael. *The Department of Mad Scientists: How DARPA Is Remaking Our World.* New York: Smithsonian Books/Harper/HarperCollins Publishers, 2009

Benford, Gregory and Elisabeth Malartre. *Beyond Human: Living with Robots and Cyborgs.* New York: Forge, 2007.

Brooks, Rodney. *Flesh and Machines: How Robots Will Change Us.* New York: Pantheon, 2002.

Capek, Karel. *R.U.R. (Rossum's Universal Robots).* Translated by Claudia Novak-Jones. Prague: Aventinum, 1921.

Fellous, Jean-Marc and Michael A. Arib, eds. *Who Needs Emotions? The Brain Meets the Robot.* New York: Oxford University Press, 2005.

Gray, Chris Habl. *Cyborg Citizen: Politics in the Posthuman Age.* New York: Routledge, 2002.

Gutkind, Lee. *Almost Human: Making Robots Think.* New York: W. W. Norton & Company, Inc., 2006.

Hall, J. Storrs. *Beyond AI: Creating the Conscience of the Machine.* Amherst, N.Y.: Prometheus, 2007.

Hall, J. Storrs. *Nanofuture: What's Next for Nanotechnology.* Amherst, N.Y.: Prometheus, 2005.

Krishnan, Armin. *Killer Robots.* Surrey, England: Ashgate, 2009.

Kurzweil, Ray. *The Singularity Is Near: When Humans Transcend Biology.* New York: Viking, 2005.

Launius, Roger D. and Howard E. McCurdy. *Robots in Space: Technology, Evolution, and Interplanetary Travel.* Baltimore: The John Hopkins University Press, 2008.

Levy, David. *Robots Unlimited: Life in a Virtual Age.* Wellesley, Mass.: A K Peters, Ltd., 2006.

Malone, Robert. *Ultimate Robot.* London: DK, 2004.

Menzel, Peter and Faith D'Aluisio. *Robo Sapiens: Evolution of a New Species.* Cambridge, Mass.: MIT Press, 2000.

Naam, Ramez. *More Than Human: Embracing the Promise of Biological Enhancement.* New York: Broadway, 2005.

Nocks, Lisa. *The Robot: The Life Story of a Technology.* Westport, Conn. Greenwood, 2007.

Perkowitz, Sidney. *Digital People: From Bionic Humans to Androids.* Washington, D.C.: Joseph Henry Press, 2004.

Riskin, Jessica, ed. *Genesis Redux: Essays in the History and Philosophy of Artificial Life.* Chicago: University of Chicago Press, 2007.

Schodt, Frederik L. *Inside the Robot Kingdom: Japan, Mechatronics and the Coming Robotopia.* New York: Kodansha America, 1988.

Singer, P. W. *Wired for War: The Robotics Revolution and Conflict in the 21st Century.* New York: Penguin, 2009.

Wallach, Wendell and Colin Allen. *Moral Machines: Teaching Robots Right from Wrong.* New York: Oxford University Press, 2009.

Wilson, Daniel H. *How To Survive a Robot Uprising: Tips on Defending Yourself Against the Coming Rebellion.* New York: Bloomsbury, 2005.

Wood, Gaby. *Edison's Eve: A Magical History of the Quest for Mechanical Life.* New York: Knopf, 2002.

Web Sites

Readers seeking additional information pertaining to robotics may wish to consult the following Web sites, all of which were operational as of this printing.

National Aeronautics and Space Administration (NASA) Jet Propulsion Laboratory

www.jpl.nasa.gov

Founded in the 1930s by the California Institute of Technology, the Jet Propulsion Laboratory is committed to exploring Earth and space with state-of-the-art rovers, orbiters, and other types of robotic spacecraft. As of this writing, the agency is overseeing more than 20 missions. This Web site provides information and news updates on all active programs, including the Spirit and Opportunity rovers, which continue to gather information on Mars, and the orbiter Dawn, which is slated to orbit the asteroid Vesta and dwarf planet Ceres.

National Robotics Engineering Center (NREC)

www.rec.ri.cmu.edu

Operated by Carnegie Mellon University's Robotics Institute, the world's largest robotics research agency, the NREC specializes in developing cutting-edge technology and building prototypes of robots later used for government and industrial purposes. This site contains information on the NREC's history, as well as detailed descriptions of its many recent projects. The site also details the Robotics Academy initiative, a program that aims to teach youngsters about robotics.

Robotics Online

www.robotics.org

Sponsored by the Robotic Industries Association, North America's only industrial-robotics trade group, this site provides a wealth of information for companies that use—or are considering using—robots as part of their day-to-day operations. The "Beginner's Guide" and "Ask the Experts" sections are of particular import for those firms looking to make their initial forays into the world of industrial robotics. The site also features a robot-buyer's guide, job listings, events calendar, and news page.

Robots.net

www.robots.net

Selected in 2002 as one of the Internet's best science Web sites by *New Scientist* magazine, Robots.net is a clearinghouse for robot-related news as well as a forum for robot builders to discuss and share photos of their creations. The site also includes a blog and events calendar, enabling robot enthusiasts to keep track of the industry's latest developments.

Additional Periodical Articles with Abstracts

More information about robotics and related subjects can be found in the following articles. Readers interested in additional articles may consult the *Readers' Guide to Periodical Literature* and other H.W. Wilson publications.

Robo-Nation. Lee Gutkind. *The American Scholar* v. 76 pp16–17 Summer 2007.

In this piece, Lee Gutkind ponders both the potential and risks of the increasing use of robotics in society. A professor of English at the University of Pittsburgh and founder of the journal *Creative Nonfiction*, Gutkind recently authored the book *Almost Human: Making Robots Think*. Questions based on Gutkind's observations on robots are presented.

A Future With Robots. Tang Yuankai. *Beijing Review* v. 52 pp42–43 August 27, 2009.

Robots are increasingly becoming available in China, Yuankai reports. Robots displayed at the recent annual China Beijing International Hi-Tech Expo attracted considerable attention, and a store that specializes in robots opened in Beijing's Zhongguancun District in April 2009. China's first intelligent robot, known as Tami, is selling well, but the price of such robots means they're still luxury items for ordinary consumers. Robot production has become a fledgling industry in China and is given considerable attention in the 11th Five-Year Plan, providing robot manufacturers with a great opportunity.

Ready To Buy A Home Robot? Cliff Edwards. *Business Week* pp84–88+ July 19, 2004.

In Japan, South Korea, Europe, and the United States, people are relying on robotic systems for repetitive chores such as vacuuming and mowing the lawn, Edwards observes. These people are betting that in the future robots will care for the elderly and sick and help people stay in touch with loved ones over vast distances. The potential market for robotic companions is very large, with one-third of the earth's population expected to be 65 or over by 2050. The writer discusses developments in entertainment robots, appliance robots, nonmoving robots, assistive mobile robots, and humanoids.

The Defecating Duck, or, the Ambiguous Origins of Artificial Life. Jessica Riskin. *Critical Inquiry* v. 29 pp599–633 Summer 2003.

With reference to the career of Jacques Vaucanson, the writer discusses the development of the science of artificial life and intelligence in the 18th century. She focuses on two of the automatons that Vaucanson created: a mechanical duck that swallowed corn and grain and then defecated, and a mechanized flute player. Vaucanson developed his experimental approach to designing automata during a period of profound uncertainty about the validity of philosophical mechanism. The ontological question of whether natural and physiological processes were essentially mechanistic, and the accompanying epistemological question of whether philosophical mechanism was the right approach to take to comprehend the nature of life, preoccupied many intellectuals in the mid-18th century. Vaucanson's defecating duck and its companion pieces received such great attention because they dramatized two contradictory claims: that living creatures were essentially machines, and that living creatures were the antithesis of machines.

Battle Bots. Chris Jozefowicz. *Current Science* v. 92 pp4–5 September 8, 2006.

This article discusses the potential for robots possessing artificial intelligence to turn evil and attack humans. A robot is defined as any machine that can sense the environment, make a decision, and act in the real world. The author discusses unmanned ground vehicles, or robot cars; swarm robots, which behave like insects and are created by engineers working in biomimetics; and androids, which look like humans.

Test-Driving the Future. Alex Stone. *Discover* v. 28 p72 May 2007.

A robot called Morpheus may herald the future of noninvasive brain-controlled robotics, Stone reports. Rajesh Rao, professor of computer science and engineering at the University of Washington in Seattle, and colleagues have designed Morpheus to respond to electrical impulses in an individual's brain. A brain-computer interface enables Morpheus to interpret these signals and respond accordingly. The ultimate goal is to develop a range of helper robots that can assist the disabled; work alongside doctors as aides; perform dangerous tasks for soldiers, bomb squads, or firefighters; or even provide companionship.

Automation Options Abound for Retail Pharmacy. Lisa B. Samalonis. *Drug Topics* v. 149 August 22, 2005.

According to the author, Todd Brown of the school of pharmacy at Northeastern University, Boston, says that automated medicine machines are a growing trend in retail pharmacies. Brown notes that although automation has been around for some time, the cost of these machines has been decreasing and has reached the point where they are beginning to be more realistic for pharmacies to use. The benefits of such automation to pharmacists include assistance with the technical dispensing-type process that allows the pharmacist to spend more time with patient-focused activities.

Robots: Nothing to Lose But Their Chains. *The Economist* v. 387 pp90–92 June 21, 2008.

Relentless advances in the power of computing mean that the latest robots are being equipped with sophisticated systems that empower them to see, feel, move, and work together, the writer observes. Robot engineers refer to these systems as "mechatronics," the alliance of mechanics, optics, electronics, computers, and software. The week of June 14, 2008, at the biannual robotics trade fair Automatica, the message broadcast was that many more robotic devices are coming to work in offices and homes. There were four trends evident at the show in Munich, Germany: robots are rapidly becoming more responsive, cheaper, simpler to program, and safer.

With a Little Help . . . *The Economist* v. 391 pp14–15 June 6, 2009.

Part of a special section on technology, this article explains why commercially available domestic robots have a long way to go before they can perform the many tasks that make up a household. Despite the fact that only a limited range of single-task domestic robots is available, there are still several observations that can be made about living with them. The first is that people will keep watching them perform their task, such as vacuuming, until they trust them completely. In addition, certain accommodations must be made to get the best out of such robots, and they eventually become part of the family.

Model Behaviour. Gianmarco Veruggio. *The Engineer* v. 293 p16 April 23–May 6 2007.

The writer asserts that applications of robots will raise ethical, social, or psychological problems in some sensitive areas. It is the responsibility of scientists and technicians who work in the field and understand the technology to address these social and ethical problems. Guidelines for the ethical application of robotics to society should be developed in transdisciplinary and multidisciplinary discussions held by scientists and scholars of humanities. Different cultures, religions, and approaches should be considered.

E-Gang: The Robots are Coming. *Forbes* v. 178 pp88–100+ September 4, 2006.

Part of a special section on technology innovators, this article discusses the work of masters of robotic innovation, entrepreneurs and researchers who combine advances in biomechanics, software, sensor technology, materials science, and computing to produce new generations of robotic assistants. These include Yoshiyuki Sankai, developer of the Hybrid Assisted Limb suit; Colin Angle and Helen Grenier, creators of the Roomba vacuum cleaner; Caleb Chung, developer of the robotic dinosaur Pleo; and Russell Taylor, creator of intelligent grippers, retractors, and other surgical tools.

Programming the Post-Human. Ellen Ullman. *Harper's* v. 305 pp60–70 October 2002.

The problem of how to distinguish human from machine is one people have brought on themselves, Ullman contends. People have conceived the idea of robots, and having dreamed them up, they feel driven to build them and to endow them with as much intelligence as they possibly can. Humans cannot resist taking up the challenge of producing tools that are smarter than they are, tools that cease in vital ways to be "theirs." Underlying that challenge is a philosophical change in the scientific view of humanity's role in the great project of life. In what could reflect supreme humility or incredible hubris, computer science has launched a debate over the coming of the "post-human": a nonbiological, sentient entity. The writer discusses the varying interpretations of life, sentience, and humanity involved in the debate.

A Job For the Droids? Tony Reichhardt. *Nature* v. 428 pp888–90 April 29, 2004.

According to Reichhardt, to achieve its objectives for human spaceflight, NASA will have to give a much higher priority to robotics. NASA invested at a moderate but steady level in general robotics research during the 1990s, with the budget peaking at $24 million per annum in 1997. However, since then, NASA's support for general robotics research has declined, mainly due to an emphasis on other technologies and financial problems in large-scale projects. Although the planetary rovers for the Mars exploration program have been an undoubted success, their mission goals are much less ambitious than those required by the new Moon-Mars program. Given this pedestrian state of the art, the goal of constructing robots to autonomously assemble a Moon base by a fixed deadline appears to be impossible. According to experts, the achievement of this objective will require NASA's funding for robotics to increase by at least an order of magnitude.

U.S. Teams Join Hands to Build Dexterous Robots. Gregory Huang. *Nature* v. 435 p861 June 16, 2005.

After years of following increasingly isolated paths, robotics researchers in the United States have agreed on the common goal of developing machines that are good with their hands, Huang observes. Robotics experts that met in Houston, Texas, in March agreed that U.S. researchers should work together to develop robots that can move around and perform useful work, a field called autonomous mobile manipulation. The idea of pooling resources was also embraced by delegates at the first annual "Robotics: Science and Systems" conference, held last week at the Massachusetts Institute of Technology (MIT) in Cambridge. Researchers hope that creating a unified scientific front will help them to compete against groups in Asia, where research into humanoid robots has been heavily funded.

Working Out the Bugs. Alison Abbott. *Nature* v. 445 pp250–53 January 18, 2007.

Studies of insect movements could help improve robot design, Abbott reports. Basic movement programs that are under the control of insects' nerve cords—equivalent to the vertebrate spinal cord—are well studied and have already been transferred to robots. Current algorithms, however, do not provide robots with the level of sophisticated and autonomous decision making that originates in the insect brain. Insect biologists have begun to decipher which particular region of an insect's brain is involved in directing different types of movement and have turned to robotics experts to test biological hypotheses about which neural networks an insect uses to navigate. In turn, the algorithms that biologists use to direct their bio-robots could one day help design more intelligent robots with the maneuverability of insects.

I Ropebot. Noel Sharkey. *New Scientist* v. 195 pp32–35 July 7, 2007.

The writer looks back through history to find the earliest programmable robots. In 1515, Leonardo da Vinci constructed a mechanical lion that was powered by clockwork. However, such devices can be traced back even further—to the Greek engineer Hero, who constructed a robot that predates da Vinci's by 1,500 years. Hero's method was unique and relied on string to program the robot's movement. Still the programmable self-propelled machine may have an even older vintage, dating to perhaps the 8th century B.C., according to references made in Homer's epic poem, *The Iliad.*

Rat-Brained Robots Take Their First Steps. Paul Marks. *New Scientist* v. 199 pp22–23 August 16, 2008.

Robots containing controls made from real neurons might provide insights into the workings of the brain, Marks observers. Researchers take isolated neurons from the cortex of a rat fetus and deposit a slim layer of them into a nutrient-rich medium on a bank of electrodes, where they start reconnecting; after about five days, patterns of electrical activity can be detected. Mark Hammond and colleagues at the University of Reading, in the United Kingdom, are trying to harness these spontaneous electrical patterns to control a robot. They hope to stimulate the neurons with signals from sensors on the robot and use the neurons' response to get the robot to respond, which could offer insights into how the brain functions and perhaps help in the treatment of disorders such as Alzheimer's disease. Meanwhile, Steve Potter of the Georgia Institute of Technology, Atlanta, is working out how to train neurons into a more permanent state of reacting to sensor inputs at the right times.

What Puts the Creepy into Robot Crawlies? Jim Giles. *New Scientist* v. 196 p32 October 27, 2007.

Researchers have elucidated the brain mechanisms responsible for eliciting uncanny feelings in people observing humanlike robots, Giles notes. Thierry Chaminade and Ayse Saygin of University College London, in Great Britain, scanned the brains of subjects watching videotapes of a lifelike robot, a less realistic robot, and a human performing the same action. They found that the lifelike robot stimulated extra activity in mirror neurons, which are active when a person imagines performing an action he is observing, possibly because the robot's movements jar with its appearance. Karl Dorman of Indiana University, has also found that robots that prompt feelings of uncanniness also tend to provoke fear, shock, disgust, and nervousness, which are typical responses to diseased bodies, suggesting that the uncanny feelings elicited by lifelike robots may represent an evolved mechanism for avoiding pathogens.

A Passion to Build a Better Robot, One With Social Skills and a Smile. Claudia Dreifus. *The New York Times* pF3 June 10, 2003.

In this article, Dreifus interviews Cynthia Breazeal, a robot designer at the Massachusetts Institute of Technology (MIT). Robotic flowers that react when a human hand is near and a robot designed for face-to-face social interaction with humans are among the innovative machines crafted by Breazeal, who describes her work in her book *Designing Sociable Robots*. Breazeal answers questions about her passion for robots and about robot design.

Yes, They Can! No, They Can't: Charges Fly in Nanobot Debate. Kenneth Chang. *The New York Times* pF3 December 9, 2003.

A nanotechnology visionary and a Nobel laureate have clashed in a debate about nanobot construction, Chang writes. The debate between K. Eric Drexler, an advocate of nanobot technology and chairman of the Foresight Institute, Palo Alto, California, and Richard E. Smalley, a nanobot skeptic and Nobel Prize-winning professor of chemistry at Rice University, Houston, Texas, began in 2001, and its latest round was published in an issue of *Chemical & Engineering News*. Drexler proposes that nanotechnology will engender nanobots that will be able to build almost anything and perform tasks like breaking down pollutants, repairing damaged cells, and even reversing the effects of aging. Smalley, however, contends that although the broader field of nanotechnology holds great promise, Drexler's nanobot vision is practically impossible. The debate has caught widespread attention among nanotechnology researchers but does not appear to have swayed many opinions.

Real World Robots. Brad Stone. *Newsweek* v. 141 pp42–44 March 24, 2003.

In recent years, robots have infiltrated the ranks of humans, although they look nothing like the luminescent-eyed androids of science-fiction stories, Stone contends. The robots are unable to emulate the human brain's limitless flexibility, but they do take advantage of the newest innovations in computing power, sensors, and artificial intelligences and can perform one or two tasks well. At present, robots are at work in homes, hospitals, and dirty, hazardous environments, such as the tunnels beneath the streets of New York City. Perhaps most important, they populate military bases across the globe, where the next generation of unmanned aerial and ground vehicles are undergoing battle testing.

Could Robot Aliens Exist? Steven Dick. *Popular Science* v. 273 p83 September 2008.

Dick considers an interesting question regarding whether robot aliens could theoretically exist. It is possible that alien beings may have already reached a point in their evolution where they have taken the next logical step and opted for robotic brains equipped with artificial intelligence. Some scientists speculate that an event called the technological singularity will occur in a few decades, in which machines possessing computer brains will become sentient and surpass human intelligence. Such an event may have already occurred in civilizations where technology is light-years ahead of that on Earth.

Robo Ronaldos. Eric Mika. *Popular Science* v. 271 p50 July 2007.

The 11th annual world cup for robots, RoboCup, was held at the Georgia Tech campus in Atlanta, Mika observes. The RoboCup tournament aims to drive innovation in robotics and to ultimately produce a team of robots capable of competing against the human World Cup soccer champions by 2050. RoboCup 2007 at Georgia Tech, which was expected to draw 358 teams from more than 20 countries, featured the Nanogram League, a tournament intended to showcase recent developments in nanotechnology. The writer also includes brief profiles of four tournament entrants.

Making Machines that Make Others of Their Kind. Adrian Cho. *Science* v. 318 p1083 November 16, 2007.

In this selection, part of a special section on robots and robotics, Cho reports that a handful of researchers are beginning to take some small steps toward the goal of creating self-replicating robots. Often seen as the stuff of nightmares in science fiction, self-replication has been a dream of roboticists for decades, and the first rigorous theory of self-replication was developed in the 1940s and 1950s by the mathematician John von Neumann. His theory, which bears striking resemblances to cell replication, was eagerly embraced by computer scientists, who have designed myriad self-replicating programs, but it does not account for the difficulties that a machine would have in gathering the required parts. In recent years, some researchers have developed simple robots that can make others like themselves from a few relatively simple parts, while others have been examining the concept of self-replication itself, arguing that there may be different levels of self-replication.

Long-Distance Robots. Mark Alpert. *Scientific American* v. 285 pp94–95 December 2001.

Alpert reports that telepresence technology has used the vast information-carrying capacity of the Internet to produce robots that can replace humans for various tasks. iRobot of Somerville, Massachusetts, has developed a telepresence robot that can travel nearly anywhere and train cameras on whatever the user wishes to see. The robot, Cobalt 2, rides on six wheels and has a pair of flippers that can extend forward for climbing stairs. The video footage, as seen through the eyes of the robot, is grainy and jerky, a trade-off for the ability to move and control the camera. iRobot also plans to market other robots with variations for business, the care industry, and defense.

Should We Make Way for the Robots and Leave the Lab? Jesse Whittock. *Times Higher Education Supplement* p20 May 7–13, 2009.

The writer considers the serious questions for the future application of academic research raised by the recent success of a fully autonomous cybernetic robot in carrying out a research project. These questions relate to whether robots could replace academics, what the ethical implications of research conducted by robots are, and whether the research project was merely a gimmick that is unlikely to be replicated in the near future.

Lights. Camera. Robot Action! Nancy Shute. *U.S. News & World Report* v. 140 pp62–63 January 15, 2006.

Paul Preston has developed a labor-and-delivery rehearsal that uses baby and mother dummies, Shute notes. Most doctors and nurses still learn their crafts the traditional way by practicing on their patients, but pressure has risen to improve patient safety. Preston, who pioneered Kaiser Permanente hospital's three-year-old perinatal safety program, has designed a robot theater as a tool for harried doctors and nurses to learn how to react with calm and precision in emergencies. Robot mothers and babies are computer controlled to simulate real obstetric emergencies at the Walnut Creek, California-based hospital.

Index

About the Editor

Writer, editor, and proponent of robot technology—particularly if it might one day save him from having to clean his own bathroom—**KENNETH PARTRIDGE** has been with the H.W. Wilson Company since 2007. He spent a year writing for the company's *Current Biography* publication before joining the General Reference department, where he's edited three volumes of The Reference Shelf. Partridge, a Connecticut native and Boston University graduate, is also a freelance journalist, and he's written about rock and pop music for the *Hartford Courant*, *The Village Voice*, *USA Today,* and AOL Music, among other publications. He lives with his wife in Brooklyn, New York.

EDISON TWP. FREE PUBLIC LIBRARY
3 9360 00671238 7